Traqueros

Traqueros

Mexican Railroad Workers in the United States 1870 to 1930

by

Jeffrey Marcos Garcilazo

foreword by
Vicki L. Ruiz

Number 6 in the Al Filo: Mexican American Studies Series
Roberto R. Calderón, Series Editor

UNT Press

Denton, Texas

10 9 8 7 6 5 4 3 2

Permissions:

University of North Texas Press
1155 Union Circle #311336
Denton, TX 76203-5017

The paper used in this book meets the minimum requirements of the American National Standard for Permanence of Paper for Printed Library Materials, z39.48.1984. Binding materials have been chosen for durability.

Library of Congress Cataloging-in-Publication Data

Garcilazo, Jeffrey Marcos, 1956-2001.
Traqueros : Mexican railroad workers in the United States, 1870 to 1930 by Jeffrey Marcos Garcilazo. — 1st ed.
p. cm. (Number 6 in the Al filo: Mexican American studies series)
Originally presented as: Thesis (Ph. D.)—University of California, Santa Barbara, 1995. Includes bibliographical references and index.
ISBN 978-1-57441-464-6 (cloth : alk. paper)
ISBN-13: 978-1-57441-627-5 (Pbk.)
ISBN 978-1-57441-480-6 (ebook)
1. Foreign workers, Mexican—Southwestern States—History. 2. Railroad construction workers—Southwestern States—History. I. Title. II. Series: Al Filo ; no. 6.

E184.M5G383 2012
331.6'27207909034—dc23 2012040679

Traqueros: Mexican Railroad Workers in the United States, 1870–1930 is Number 6 in the Al Filo: Mexican American Studies Series

This publication was supported by a generous gift from Alex and Toni Garcilazo.

contents

List of Tables and Maps

Tables

Map

foreword

OVER A CENTURY AGO RAILROADS DOMINATED THE ECONOMIC and physical landscape of the United States with 93,000 miles of track crisscrossing the nation. Who laid the tracks? Who kept the trains running, the lifeblood of national economic development? Railroad workers represented a diverse panorama including the Chinese, European immigrants, and African Americans. Mexican rail hands, however, have remained unacknowledged. Yet, at the dawn of the twentieth-century, they laid over fifty percent of the tracks for western lines. *Traqueros: Mexican Railroad Workers in the United States, 1870–1930* is their story, ably told by an accomplished historian, the late Jeffrey Garcilazo.

Garcilazo came to this study not as an inquisitive graduate student hunting an interesting dissertation topic, but as a man who had come of age in Colton, a southern California rail town and who, as a young adult, had organized with the United Farm Workers. His passion for Chicano history was borne out of his own lived experiences. Emphasizing collective action over individual efforts, he understood the meaning of community, as practice and principle. Furthermore, Jeffrey Garcilazo understood the power of history to inspire students, a fact not lost upon current Arizona politicians and school board members who have denied Mexican American children access to their history.[1] Indeed, *Traqueros* will take its place in the canons of Chicano history and labor studies, but more importantly, it will find its way into the hands of youth eager to see themselves in American history books. A vibrant narrative, *Traqueros* accentuates the enduring relevance of social history to connect our past to our present.

This work stems from Jeffrey Garcilazo's dissertation that he completed at UC Santa Barbara in 1993 and he was in the process of revising it for publication when he met his untimely death. As the first comprehensive treatment of Mexican railroad workers, this book forges new ground in several areas. First, next to agriculture, the railroad industry in the early decades of the twentieth century was the major employer of Mexican immigrant labor. Railroads regularly sent recruiters to the border and even into the interior of Mexico to enlist men for track labor and maintenance work, offering such incentives as transportation and housing for themselves and their families. Second, the bulk of Chicano historiography has focused on people and events in the Southwest and the few books on the Midwest tend to focus on urban areas such as Chicago and Detroit.[2] By focusing on the boxcar barrios in places like Argentine, Kansas, Garcilazo widens the lens of Mexican American history in interesting, meaningful ways. Unlike many traditional labor histories, which focus on men on the factory floor or in the fields, Jeffrey Garcilazo emphasizes the integral nature of family settlement in shaping working-class communities and consciousness. With men at work on the tracks, women made boxcars their homes, planting gardens, hanging curtains, and caring for children. Many reached out to one another as *comadres* sharing confidences and pooling resources. Frederica Vásquez remembered the excitement surrounding the arrival of a new crew train: "*las señoras* went out to meet them and brought them food and brought them clothing and made them feel very welcome."[3] While these makeshift settlements were hardly working-class utopias—far from it—they nourished a sense of belonging as *traqueros* and *mexicanos*.

Jeffrey Garcilazo, moreover, makes the connections between the migratory pattern of Mexican railroad workers and the development of small town barrios in the Midwest. By focusing on Kansas, he explores the creation and persistence of Mexican ethnic communities, which began as boxcar settlements. Yet, as he points out, Mexican families did not live in a hermetic environment, sealed off from American cultural influences. Like their urban peers, Mexican women and children became a prime audience for Progressive reformers, with the Young Men's Christian Association (YMCA) tailoring its Americanization programs specifically for *traquero* families throughout the rural Midwest.

A meticulous researcher, Garcilazo has gathered a stunning array of archival materials on Mexican railroad workers. From his sources, he reconstructs episodes of daily life from lonely encampments to spontaneous strikes. His arguments on economic stratification, ethnic enclaves, family ties, and identity are solidly grounded in primary documents. Never overstepping the bounds of his evidence, Garcilazo situates *traqueros* within the larger context of American working-class history and his keen insights have retained their currency over the passage of time. Scholars familiar with his research have long anticipated publication of this manuscript. In the words of political theorist Rodolfo Torres: "When one of the late labor historians in the United States brings his thinking to bear on the meaning of labor and class, the result is this brilliant book that shows another world is possible." In a heartfelt, more understated message Latin American historian Steven Topik remarked: "It was such a tragic loss, not only because we missed his friendship...but also missed his dedicated scholarship. I am glad that The University of North Texas Press is letting a wider public get a taste of what Jeff gave us and what we have lost." I would like to thank Roberto Calderón and Ronald Chrisman for their steadfast support and stewardship. Without their leadership and corazón, this important contribution would have remained in abeyance. If he had lived, Jeff would have been proud to place his first academic book with the Al Filo: Mexican American Studies series at the University of North Texas Press.

Traqueros became Jeffrey Garcilazo's life work and its publication a fitting legacy for a historian remembered for his big personality and even bigger heart. For Jeff, teaching was never a job, but a calling. He devoted untold hours to students—encouraging, prodding, and challenging. In the words of his *compadre* Gabriel Gutiérrez, a professor of Chicana and Chicano Studies at Cal State Northridge: "When I became a graduate student, my interactions and sometimes debates with Jeff led to [the] realization that there was no finality in 'conclusions,' that the arrival at such was merely a point of departure for another journey." One colleague, who was in graduate school with Jeff, recalled how this "proud Chicano" once queried her about her name, Rachel. "He then turned to me and said, 'Why don't you go by Raquél?' What he was really saying...was who are you...what do you stand for...who do you want to be? At that time I was extremely shy...something Jeff understood, but he knew if I wanted to be a professor I [would] need more confidence and strength." Recounting her epiphany, she declared, "I went home and thought about it and it was at that moment...I became María Raquél Casas, Chicana historian."

At a dissertation workshop in July 1990, I was assigned as Jeff's faculty mentor. At dinner the evening before I arrived, he purportedly made quite an impression on fellow participants. A local hero, Jeff Garcilazo, in his best Colton homeboy persona, had defended a young colleague from the unwanted attentions of a nearby diner. More a coach than a mentor, I greatly enjoyed working with Jeff as we went over his chapter drafts. We stayed in touch throughout the writing process and beyond. Always generous, he invited me to UC Irvine in April 1998 to give a talk on my new book *From Out of the Shadows*. Later that year, he had a routine surgery that went terribly wrong and he slipped into a coma for several weeks. Gabriel Gutiérrez recalled "visiting him weekly in the hospital, conversing with his parents...occasionally taking part in his physical therapy sessions once he awoke from the coma, being witness to the determination he showed...even though his speech was limited...he had moments when he was lucid and even cracked jokes, to the day I received a call from his father informing me 'We lost him.'" Jeffrey Garcilazo died on May 4, 2001.

Reflecting Jeffrey Garcilazo's imprint as an educator and scholar, his former doctoral students Enrique Buelna and Steven Rosales have launched their own academic careers, at Cabrillo College and Grand Valley State, respectively. Distinguished historian Mario García, Jeff's own advisor, considered him "one of the best graduate students I had the fortune to work with over the years." He further reflected, "I miss Jeff and the profession lost a wonderful historian and I am pleased and thankful that his groundbreaking [work] on Mexican immigrant workers is being published."

In this brief foreword, I have strived to balance the strength of the scholarship with the measure of the man. *Traqueros: Mexican Railroad Workers in the United States, 1870–1930* is Jeffrey Garcilazo's intellectual gift. Labor historian Heidi Tinsman perhaps said it best: "Jeff brings to life the world of the boxcar communities and migrant laborers that stretched from San Bernardino to Pennsylvania. Tenderly...he recovers their longing, sadness, humor, and fight." Their history unfolds on the following pages.

Introduction

Former *traquero* Jesús Ramírez complained that "It is never in the books or papers that the Mexicans built the railroads. And we had no machines, only our hands."[1] It must also be stated that even Chicano historians have never included Mexican women as railroad workers, if not necessarily at work on the track, then as workers at home in boxcars and railroad yard shanties. In order to at least partially fill this gap in labor and Chicano and Chicana history, this study will show that the development and socioeconomic growth of the Southwest were directly tied to the toil and sacrifice of tens of thousands of workers of Mexican origin,[2] without whom the railroads in the West and Southwest may not have been built when they were, or as quickly.[3]

While the books and the newspapers rarely tell the story of how Mexicans "built the railroad," those of the Immigrant Generation possess the collective memory of how they traveled from their homes in Mexico to various destinations in the United States, the vast majority of whom either worked on the railroad or traveled on the railroads to other employment sectors. This study is an attempt to reconstruct their daily lives and breathe life into scattered pieces of historical materials and raw data that remain available. In so doing, it will first discuss the pertinent literature in railroad, labor, and Chicano history, and show how others have long recognized the important contribution of Mexican workers on the railroads, but failed to place Mexicans at center stage in the development of the railroad into the Southwest and West.

The present study also builds upon the "political generational" framework recently articulated by Mario García.[4] A political generation is not a biological generation. However, a political generation may parallel a biological generation and theoretically span more than one biological generation. The most salient feature of a political generation is a set of common life experiences, such as wars, economic depressions, and mass migrations. García has termed the generation of around one million Mexicans who sojourned north of the border between 1900 and 1930 the Immigrant Generation because of their common experience as "immigrants." And, while the process of immigration preceded 1900 and continued after 1930, the percentage of Mexican immigrants in United States communities is unparalleled.

Traqueros falls under the rubric of the "new social history" and "new labor history." As such, it seeks to examine the intersections of race/culture, class, and gender among the Mexican railroad worker communities in the United States. Like many other labor studies that have been influenced by the works of historians such as E. P. Thompson and Herbert Gutman, this study seeks to understand not only what workers did while at work and the associations they formed, but also what they did at home in their leisure time, and how they both accommodated and resisted cultural assimilation. Ultimately, this study like other ethnic and labor histories seeks to understand Mexican railroad worker culture, that is, how Mexican railroad workers understood themselves and how they negotiated their lives around work and leisure on a daily basis.

As the title indicates, this study is largely limited to those Mexican railroad workers north of the Rio Grande. However, there are clear linkages between Mexican railroad workers' experiences in Mexico and their search for employment on American railroads. Certainly, northern Mexico and the southwestern part of the United States can be understood as a unitary socioeconomic region, yet the differences may be far greater than the similarities.

Mexican track workers, like other track workers, were unfairly and inaccurately categorized by the railroads and mining industries as "casual" or "common" labor. These terms are euphemisms for "unskilled" or "semi-skilled" workers, which when applied to immigrant workers, especially Chinese, Irish, Greeks, Italians, Indians and Mexicans, implied that their work could be performed by any strong able-bodied man. Defined this way, some historians have also denied Mexican workers the dignity and recognition for having built much of the railway lines in the Southwest and Central Plains region. The exact mileage of track laid and maintained by Mexican workers compared to that of Euro-American groups is ultimately less important than the fact that each group of workers gradually replaced the previous one in successive waves as "common laborers." It is virtually impossible to determine with any great degree of accuracy which workers built how much of any one railroad since these groups worked on the railroads at the same time, often on opposite sides of the tracks.[5]

The older literature reflected and reproduced a narrow view of history. For instance, one important historian of the Santa Fe railroad only backhandedly acknowledged the contributions of Mexican track workers. In his book, *Steel Rails to Santa Fe* (1950) L. L. Waters rationalized his omission of Mexican workers for two reasons. First, because "they [Mexican workers] were not, in a strict sense, employees" of the Santa Fe.[6] Instead, Mexican railroad workers had been hired by labor contracting agencies and not by the railroad company itself. Secondly, Waters believed that Mexican workers were not "permanently identified" with the railroad and "were more or less transient and *casual laborers* [emphasis added]." The fact that "casual laborers" by definition were (and continue to be) transient workers, does not diminish their importance as actors in the construction and maintenance of the railroad.

Other railroad historians mention Mexican and American Indian track workers only in passing, or do so pejoratively.[7] This rationale for the exclusion of tens of thousands of Mexican workers from the recorded history of the Santa Fe railroad underscores the invisibility of "casual labor" in American society. It would not be too far off the mark to suggest that Waters and other old-time railroad historians deliberately avoided any discussion of Mexican workers that would have shown an unromantic side of the railroad industry. On the other hand, some older but more useful studies by Paul S. Taylor and Victor S. Clark took a keen interest in the influx of Mexican workers during the first decades of the twentieth century and made important, albeit biased, observations about Mexican workers (especially Clark). While neither of these studies focuses entirely on Mexican railroad workers they are useful early studies of Mexican labor in the United States that provide important information about such workers for the railroad.[8]

Traqueros owes an intellectual debt to the scholarship of Walter Licht and David Montgomery.[9] Licht provides a useful approach to understanding railroad worker culture during the nineteenth century; however, it is largely limited to Northeastern railroads. Most significant is Montgomery's examination of "common labor" since the majority of the historical actors in the present study fall under this crude category of the American working-class. According to David Montgomery, common laborers tended to work in gangs and be male, immigrant, single, and the lowest paid.[10] However, neither study examines with any depth the experiences of Mexican railroad workers in the United States.

Perhaps the study that best examines Mexican railroad workers and community formation is an unpublished dissertation by Judith Laird, which examines the Mexican community in Argentine, Kansas.[11] This work is the most extensive examination of community formation by Mexicans in the Central Plains.[12] And while Laird does not directly focus on the work lives of Mexican railroad workers, she does examine the influence of the railroads in terms of chain migration, chain occupations, and housing. For example, she shows how Mexicans in Argentine typically came from the same villages in the Mexican states of Michoacan and Guanajuato. These workers used their social networks—relatives, *compadres* (godparents), neighbors, and friends—to find jobs on specific railroads, divisions, sections, and work gangs. Of course Chicano historians and social scientists have also long identified these processes and added to our overall knowledge of the Mexican immigrant/migrant experience. Michael Smith has also contributed much to our knowledge about Mexican immigrants in the Central Plains. While he relies heavily on Judith Laird, he argues that Chicano history must be expanded to include areas beyond the Southwest. And like others, he discusses railroad work as backdrop to community formation, immigration and acculturation.

Chicano historians interested in telling the story of community formation and ethnic identity have also trod an important path of investigation for the current study. However, not unlike other historians, they have only indirectly touched on the significance of the railroad as a structural institution which shaped the work lives of Mexican immigrants/migrants. Moreover, even fewer have identified railroad employment, especially track work, as the first significant industrial type work ever performed by *Hispano* and Mexican workers in the annexed Southwest.

At the beginning and throughout the process of southwestern and United States economic development, Mexican workers were inextricably linked to the railroads, mining, ranching and agriculture. But the railroad was the physical conduit in which Mexican workers found both employment and transportation to virtually every section of the country. The physical magnitude and bureaucratic complexity of the railroad, "the nation's first big business" both in the United States and Mexico, made it the largest and often the first employer of Mexican labor in the United States. Indeed, the immensity and impersonal corporate management of the railroads must have overwhelmed most immigrant Mexican workers upon their initial encounter. Certainly in the minds of working-class *mexicanos*, the "colossus of the north" was best symbolized by the regular passing of the noisy and earth-shaking power of the locomotives and boxcars. Moreover, the sheer multitude of people employed by the railroads probably represented the ideal opportunity, at first, for stable employment and steady wages for new arrivals from Mexico.[13]

Chapter One discusses the social and economic growth of the Southwest and northern Mexico, demonstrating the linkage between the United States and Mexican railroads and the centrality of Mexican labor therein. It also shows how the railroads gave rise to a dynamic regional economy based on Mexican labor and eastern industrial centers and markets. Chapter Two examines the origins, processes, and mechanisms of Mexican labor recruitment and immigration and discusses the workers' ability to determine the conditions of their work. A description of track work is provided in Chapter Three. It specifically delineates the areas of contestation (between workers and foremen), accidents and subaltern forms of resistance.[14] Chapter Four reconstructs the conditions under which Mexican track workers formally resisted exploitation and investigates the variety of conditions under which traqueros organized themselves and engaged in strike activity. An examination of the origins and formation of Mexican boxcar communities throughout the Midwest, Plains, and the Southwest is offered in Chapter Five. Finally, Chapter Six discusses "traquero culture," which is the unique conditions and consciousness that emerged specifically from the daily life of Mexicans who lived and worked on and near the railroad roundhouse.

Chapter 1

Railroads and the Socioeconomic Development of the Southwest

In popular lore of the American West, only Chinese and Irish workers built the railroads, laying track, digging tunnels, and building trestles and bridges. Indeed, this picture of track work in the West is true for the transcontinental railroad of 1869, but not for the decades to follow. During the late 1800s, virtually all types of native-born and immigrant labor worked on the tracks in this region at one time or another. However, by the turn of the century, Mexican immigrant labor far outnumbered all other groups of immigrant and or native-born labor on the tracks in the Southwest. In order to understand the twentieth-century experience of Jesús Ramírez, whose words opened this study, and of the hundreds of thousands of other people of Mexican background who found work on American railroads, we must first understand the prevailing social and economic conditions under which Mexicans lived during the late nineteenth and early twentieth century.

The U.S.-Mexican War (1846–1848) and the annexation by the United States of Mexico's northern borderlands shaped the socioeconomic relations between two peoples and cultures. The impact of this war tossed people of Mexican background into a similar social trajectory in the United States as other non-white peoples especially American Indians and Africans. Ultimately, this chapter also discusses the emergence and impact of the railroads on the development of the Southwest. The present study borrows from and complements other studies that deal with the relationship between core and peripheral economies. In so doing, this study aims lower. Indeed, it examines the base of this relationship: the relationship between non-white conquered Mexicans (and immigrants) and the larger issues of working-class and labor history in North America. Finally, this chapter explains the social transformation of ethnic track labor and why by the 1880s track labor became synonymous with racial subordination.

Long before the Mexican immigrant diaspora and before the first railroad spike was driven anywhere in North America, the American Indian and mestizo people of the Mexican borderlands were the travelers and traders of the region (1492–1821). Before them, people whom some anthropologists call the Anasazi (800–1492), crisscrossed northern Mexico (today's Southwest) and Great Plains region for trade and to hunt and sometimes battle. As a frontier region, Athapascan Indians, Spanish colonists, Tlaxcalan Indians and mestizos from central Mexico populated northern Mexico and shared the colonial legacy of mestizaje. Indeed, the Europeans and racial mixes added to the already diverse indigenous inhabitants. Speaking of the nineteenth century, Deena González argues, "Biculturalism evolved throughout Mexico's frontiers into a complex system of communication based on more than the blending of races. Santa Fe [was] built on the legacy of racial mixing and cultural borrowing while conforming as well to another aged pattern of inter-Mexican migration."[1]

The indigenous routes of trade that crisscrossed the region during the pre-colonial period are perhaps the most significant legacy of the Indian people of the Southwest. For instance the Atchison, Topeka and Santa Fe (AT&SF) Railroad built the line that connected the United States with Mexico along the Rio Grande (Rio Bravo to Mexicans), the major artery of trade originally established by indigenous peoples and later used by Spanish and Mexican travelers and traders. During the Mexican period, the Chihuahua and the Santa Fe trade expanded the pre-colonial trading patterns in this area and connected Santa Fe, New Mexico with Ciudad Chihuahua and St. Louis, Missouri. The Santa Fe Railroad also laid track roughly along the same route as the old Santa Fe Trail. Thus the extension of American railroads into the Southwest does not represent a total break with the past, but rather a form of both continuity and change.[2]

Historian Mario García sums up the significance best in his important book on the Mexicans of El Paso, *Desert Immigrants*: "Although the railroad brought modernized transportation, it did not represent a total break with the past. The railroads were built on the Spanish and Mexican trails that crisscrossed the Southwest and made El Paso a commercial stop on the Old Spanish Trail."[3] In addition, the pre-colonial indigenous trade routes preceded the Spanish and Mexican periods. El Paso historian, Owen White, possessed a different view of Mexican influence on the railroad. White wrote, "El Paso got the railroads with their shops and their payrolls because the Americans in the town went after the business, while the Mexicans of Ysleta and Socorro sat around following the shade from one side of the house to the other."[4] Such images tinged with Manifest Destiny die hard.

The U.S.-Mexican War (1846–1848) marked the early manifestation of nineteenth-century Anglo notions of racial superiority and imperial greed. The invasion of Mexico's northern territory and the conquest and invasion of Mexico City itself, resulted in the annexation of slightly more than half of Mexico's land and the imposition of American culture and laws on the people of the Southwest. After military conquest of both the Southwest and Mexico City, in exchange for the paltry sum of fifteen million dollars and an additional sum for reparations, the United States gained what is now the states of Texas, New Mexico, southern Colorado, Utah, Arizona, Nevada, and California. In 1849 alone, three times this amount was recovered in gold from California.[5]

In the minds of most Euro Americans, their victory over Mexico reinforced the notion of white supremacy and Manifest Destiny. This conquest opened the Southwest to American technology and markets, which in turn made Mexican labor in the newly conquered lands available for appropriation.

For Mexicans, the experience of conquest and subordination shifted and recreated new patterns of social relations between Mexicans and Anglos throughout the nineteenth and twentieth century. Of course, it must be quickly added that some Mexicans, especially *los ricos*, the frontier elite (who usually self-identified as Spanish), experienced the conquest differently. *Los ricos* had more to lose in the form of land and personal wealth. Some collaborated with Anglos in order to retain their wealth and privilege. Late nineteenth-century immigrants, on the other hand, did not experience the conquest themselves although they certainly had knowledge of it; therefore, Mexican immigration after 1880 represents a major watershed in the political makeup of Mexican origin population in the United States.

The Treaty of Guadalupe Hidalgo promised among other things that Mexican rights to property, culture, and customs would be respected. Original inhabitants had one year to decide to leave the annexed territory or remain in their homes and automatically receive American citizenship with full rights and privileges. Unfortunately for the vast majority of Mexicans, even the self-identified Spanish *gente de razón* (people of reason or status), eventually found themselves subjected to racial/ethnic discrimination, political manipulation, and cultural alienation. In 1853, the Gadsden Purchase placed the region of southern Arizona within the boundaries of the United States for the ostensible purpose of direct railway linkage with California. However, a more compelling motive for the Gadsden Purchase lay in the potential profits of mine operators. In 1924, one historian observed: "That sixty mile drop in the line gave the United States the control of the richest Copper deposits in the world..."[6] The discovery of copper and other metals in southern Arizona later yielded high profits for Eastern capitalists.

This pattern of land loss and alienation has been discussed elsewhere, but the following passage from a New Mexican newspaper in 1881 is typical of the kind of attitude and activity that effectively alienated Hispanos and Mexicans from wealth: "A Mexican has found a new coal vein a few miles out from Las Vegas in [the] direction of Los Alamos. Some lively Yankee should hunt up the outcrop and lay claim to the location."[7] And while the war with the United States marked an important watershed for Mexico and its people, social inequality and cultural institutions continued into the second half of the nineteenth century.

Ironically, while railroads often laid track upon the former Mexican trade routes, railroad penetration into the West marked the end of an era for other indigenous peoples, the Plains Indians. In 1883, General William Sherman stated, "These roads enable us to send soldiers to threatened points at the rate of five hundred miles a day, thus overcoming the space in one day which used to require a full month of painful marching."[8] At the end of his military career, he further claimed:

> A vast domain, equal to two-thirds of the whole United States, has been thus made accessible to the immigrant, and in a military sense, our troops may be assembled at strategic points, and sent promptly to the place of disturbance, checking disorders in the bud.[9]

Railroads routinely provided arms for its track crews in order to quell Indian resistance. In 1866, Chief Red Cloud, of the Sioux nation, personally warned railroad engineers in Wyoming to turn and leave. He said, "We do not want you here. You are scaring away the buffalo." To preserve their culture and stall the encroachment of the "bad medicine wagon," Indian bands attacked track crews and cut telegraph lines.[10] Some white men mistakenly thought that Indians "had a mortal fear of telegraph wires and rails. They would follow the road for miles if need be to find a bridge where they could cross under, for nothing would induce them to cross the track at grade."[11] Of course, given the rate at which the military killed off and relocated Indian tribes, the Indian fear of railroads is understandable.

The idea of "mission" resonated widely among Euro Americans who rationalized the demise of indigenous cultures. In 1911 a writer for *Santa Fe Magazine* summed up the importance of this "mission" of the Santa Fe Railroad: "its mission has been to supplement the historical achievements of Houston, Kearney, Carson, Stockton, Fremont, and Custer, and to cooperate with the true American spirit in building an industrial empire upon the foundations of military conquest."[12] Eighty years later the historian Ronald Takaki offered a markedly different assessment: "The railroad and the forces it represented—the expansion of white settlement, civilization, and the market—made it apparent ... that the Indian of the 'past' had no place in technological America."[13]

Mexicans and Indians represented obstacles to the railroad and America's hegemony over the West. The official justification for the United States Congress to support the building of the first transcontinental railroad consisted of:

> (1) That it was a political necessity, and would prevent the loss to the Union of the Pacific States; (2) That it was a military necessity, and would enable the Government, by the rapid movement of troops, to resist the invasion of a foreign enemy; (3) That it would end the Indian wars; (4) That it would furnish a cheaper and more rapid means of transportation for mails, troops and munitions of war; and (5) That it would lead to the development of the resources of the vast and then unpeopled territory between the Missouri and Sacramento Rivers.[14]

Indeed, with the majority of the population of the West Coast "consisting chiefly of Mexicans," Congress expressed concern about the possibility that Mexicans would retake the conquered lands. Politically, the railroad reinforced the solidarity of the loyal states of the Union and, according to historian Yen Tzu-Kuei, some government officials feared that strong Southern sentiment in California "might be used to place the new state in the Confederacy or to bring about its return to Mexico."[15] Hardly a generation had passed since the end of the U.S.-Mexican War and Congress feared Mexican insurrection in the conquered lands. Thus, the extension of the railroad and the ability to move troops into the West was almost a necessary prerequisite for white settlement and Anglo American hegemony.[16]

While the completion of the first transcontinental railroad (1869) was certainly a magnificent feat of engineering and labor—especially by the Chinese who built east across the Sierras and the Irish who built west from Omaha—it symbolized for many the end of the frontier. Indeed, for the first ten years of its life it was more a symbol of American industrialization than it was completely operational and commercially profitable. Hastily laid rails and trestles that swayed in the wind made heavy use unsafe. Among other obstacles, winter snow drifts proved especially halting to the frequency of successful train shipments. According to one historian of Chinese labor, "Westbound railroad freight, even in the peak years of 1880, never equaled one-third of the tonnage of import by sea."[17] Until the construction of the Southern Pacific and the Santa Fe Railroads in the Southwest, mining, agriculture, and ranching production plus extensive trade between the East and the West coast was limited.

The notion of the Southwest as boundless untouched country resonated widely throughout the United States. And even though the Census Bureau officially declared the frontier filled up by 1890, newspapers and railroads in the Midwest encouraged entrepreneurial white settlers to build their homes and fortunes in the Southwest. For example, in 1881, the *Silver City Chronicle* summed up the impact of the railroad on the West:

> When the railroad passed thru Utah it sounded the death knell of polygamy; ever since it has been dying, slowly dying. When the M. K. & T. [Missouri, Kansas & Texas] road was allowed to cut in twain the five Indian nations to our east, the red inhabitant could no longer bar the doors against his pale

> faced brother. When the A. T. & S. F. road shall have passed down and along the banks of the Rio Grande from Wallace to El Paso, then the days of pointed stick plowing virtually ends in the rich valley of the Nile of America. A new era is dawning. The sharp, far-seeing Yankee is dropping down here and there, and more are coming. They will look from the car window at this rich vine-growing and grain-producing land and will possess much of the same. Enterprise backed with push is following in the wake of the Great Bonanza Civilizer.[18]

Thus by implication the *Chronicle* lumped together Mormons, Plains Indians, and the "stick plowing" inhabitants of the Rio Grande region, presumably Mexicans and Pueblo Indians, as dull, anachronistic and short-sighted decaying threats to the Great Bonanza Civilizer. In this one quote, monogamy, racial homogeneity, and technological innovation ring as the reigning ethos of the day. In 1905, twenty-four years later, the *El Paso Daily Times* echoed much the same sentiment:

> The dawn of the greater southwest is at hand. The railroads traversing the vast undeveloped sections of Texas, Oklahoma, Arizona and New Mexico, are completing preparations for one of the greatest campaigns for territory upbuilding in the history of the country.... The success of the movement will depend largely upon the willingness of desirable immigrants to move to the untouched fields for future homes and fortunes.[19]

In other words, the *Times* believed that the original inhabitants of the Southwest were unfit to bring progress.

The perception of Mexicans and things Mexican in the Southwest was, like the land, to be overturned and harnessed. The following account of a traveler aboard a Southwest bound Santa Fe train parallels the above views:

> As we rushed along, gangs of Mexicans—track-menders—in tall, pointed straw hats, which appear in the East only on the covers of cigar boxes, stepped aside to let us pass. Strange brown little birds scattered like chaff, but fell behind in their race with us.[20]

The author probably did not consciously intend to equate Mexicans with "strange brown little birds"; nonetheless this passage suggests that the writer saw both Mexican "track-menders" and "brown little birds" as failures in the "race" with Anglo American locomotive technology. These views represented the contemporary thinking about the Southwest and the West.

The period between 1790 and 1860 witnessed inimitable territorial growth, tripling the land area of the United States. By 1879, ten years after the railroads linked the West to the East, track construction totaled fifty-three thousand miles. A decade later the rail linkage of the county totaled ninety-three thousand miles of operating line and amounted to $5.4 billion. And, of course, the railroad greatly facilitated the Americanization of the West and Southwest. According to Ronald Takaki, Euro Americans living in Indian Territory between Kansas and Texas numbered about 7,000 in 1880. Nine years later, and five years after the railroad reached this area, the number of Euro Americans had dramatically increased to 110,000.[21]

The railroads provided transportation and communication links across the continent, but equally important, the railroads themselves became large consumers of steel, coal, lumber, timber, and other raw materials. As such, railroads stimulated the development of basic industries. And as America's first "Big Business," the railroads also faced the responsibility of designing modern management techniques for its employees.[22] A glimpse of the numbers who worked for the railroad compared to other industries provides an excellent picture of the immensity of the railroad. In 1880, for example, textile mill workers numbered slightly over 300,000 and the number of miners was about 230,000; meanwhile railroaders totaled well over 400,000. Additionally, according to James H. Ducker, "while many miners and textile workers labored for relatively small firms, nearly three-quarters of all railroaders worked for roads employing a thousand or more."[23] For the purpose of comparison, Alfred D. Chandler notes that "In 1891 the Pennsylvania Railroad employed over 110,000 workers. In the same year the total number of soldiers, sailors, and marines in the United States armed services was 39,492." He further notes that in 1891, the Post Office, "the largest government office in terms of employees, had 95,440 workers."[24] Thus measured, the Pennsylvania Railroad alone exemplified the growth and complexity of modern capitalist operations throughout the eastern United States, and by implication the Southwest. The East remained the original core of industry and manufacturing and made the Southwest and West dependent on eastern-based railroads, banking, and technology. The railroad served the market as a conduit for transporting raw materials, finished products, and labor. In turn railroads facilitated the dynamic growth of towns and settlements that grew as the extractive economy of the Southwest flourished. The decade beginning with 1880 marked another great period in railroad building. Walter Licht informs us that the country was linked by "ninety-three thousand miles of operating line at the cost of $5.4 billion."[25]

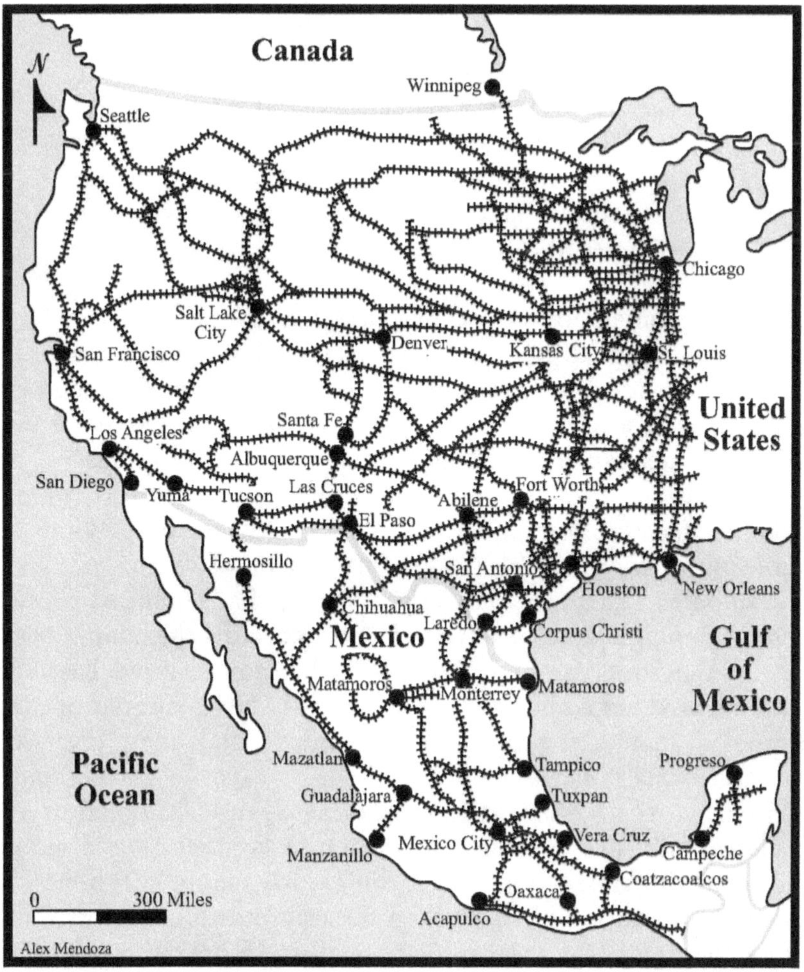

MEXICAN AND WESTERN U.S. RAILROAD NETWORK, 1920

El Paso best represents the centrality of the railroad to the development of the Southwest and the economic link to eastern markets.

As García compellingly shows, El Paso was to the Southwest what Ellis Island was to the East. While not the only terminus for immigrants coming from Mexico, El Paso was no doubt the most important border city connecting Mexico to the United States. The *Modern Traveler*, a railway passenger journal published in Chicago, had this to say about the "El Paso Route":

> Four gigantic railroads from four mighty cities, at four points of the compass enter the city of El Paso like the four spokes of a great wheel converging to the hub, covering a sweep of country unparalleled in the world for extent, wealth and magnificence. The Santa Fe comes in from far away Chicago, the Texas & Pacific from New Orleans. the [*sic*] Mexican Central from the city of Mexico and the Southern Pacific from both New Orleans and San Francisco, yes, and away up to Portland, Oregon.[26]

Another contemporary writer compared El Paso favorably to Gibraltar as a signal point of transportation and trade between the Mediterranean and the Atlantic. In 1881, *The Lone Star* accurately predicted El Paso would become the major border trade center with major rail lines converging from the four cardinal directions.[27]

The railroad dramatically transformed other sectors of the economy, specifically mining, agriculture and ranching. While the expansion of the railroad marked the decline of wagon and mule train shipment, the railroads also profoundly changed the nature and scope of mining operations. Aside from connecting the mining operations of the Southwest and West with the industrialized East, railroads allowed mine operators to transport larger quantities of earth in order to make the extraction of precious metals more profitable than previously possible.

Mining in the Southwest began during Spanish colonialism and developed only slightly under Mexican rule. And while gold mining coincided with annexation in California, early mining was limited to surface ores that could be worked through individual effort. Long before the arrival of the railroad, small-scale operators extracted copper from the Santa Rita de Cobre mine in New Mexico. In 1820, the discovery of placer gold in the Ortiz Mountains (south of Santa Fe) caused short-lived excitement. But for the most part, Americans introduced modern mining techniques during the late nineteenth century.

The combination of the demand for railroad transportation and the encouragement by railroads to invest in western mining hastened the mining boom of the 1880s.[28]

Before the costs of supplies and shipment were reduced by the introduction of the railroad, miners could only profitably work the richest mineral deposits. Mine operators were compelled to use time-consuming and expensive pack trains and wagons. According to Richard White, one man, George Hearst, "paid $30,000 in freight costs alone to get a single 80-stamp (that is, a mill with 80 stamps to crush the ore) into the Black Hills." Another example of a valuable mine (marble, lime, and cement) that was impossible to mine profitably without the railroad was Slover Mountain near the pueblo of Agua Mansa in what is now Colton, California. One newspaper noted that the "Slover Mountain ledge has been known for some time, but no effort has been made to work it on account of want of capital on the part of its owners."[29] When railroads finally arrived, the cost of transportation dropped low enough for operators to extract lower grades of ore profitably. The open-cut mining method represents one such technique first perfected at the Bingham Canyon mine in Utah.

Bingham Canyon just south of Salt Lake City was one of the largest open pit copper mines in the world. Miners referred to it as "the richest hole on earth." The Guggenheim family first developed Bingham Canyon and passed it on to the Kennecott Copper Company in 1910. Four years later Kennecott reported that the mine had already produced upwards of $1 billion worth of copper. Indeed, like other parts of the Southwest, railroads and mining companies imported thousands of Hispano workers from southern Colorado and northern New Mexico, along with European and Asian immigrants, to insure a profitable return on the ore.[30]

Almost every western state contained a great abundance of subsurface minerals. By 1900 highly mechanized mining operations controlled by Eastern corporate interests predominated in the Southwest. In 1910, the Kennecott Copper Company bought the Santa Rita (de Cobre) copper mine and transformed it into one of the largest open pit mines in the Southwest. In 1914, western miners produced 90 percent of the nation's copper, a large portion of its lead, and most of its gold and silver.[31]

Mining interests regarded copper as the "queen" mineral in the Southwest. Rapid industrialization demanded increased copper mining to supply the expanding telegraph, electrical, and manufacturing industries. By 1917, copper production in Arizona had increased to 23,274,965 pounds from 719,035,514 pounds in 1883. And by World War I, Arizona led the country in the extraction of this valuable industrial mineral. Given the subordinate position of Mexican labor in the copper mining industry, it may not be too far off the mark to suggest that mine operators perceived Mexicans as landless serfs to the "queen" of the mine metal industry.[32]

The proximity of the western copper industry to Mexico made it necessary for American capitalists to woo Mexican mining interests in modernization in order to both access the extensive mineral resources of Northern Mexico and to recruit Mexican labor. Mexican land and workers provided the backbone for mining on both sides of the border; however, skilled occupations tended to be the domain of white skilled employees imported from the East, and companies hired unskilled workers locally and imported them from Mexico proper. El Paso's importance increased in direct proportion with the development of the mining industry. In 1901, the *International Industrial Record* listed the several El Paso mining companies including the Pass City Copper Company assessed at $100,000, and La Fortuna Mine and Milling Company, valued at $199,950. Both corporations owned mines in northern Mexico.[33]

Southwest mining and smelting companies followed the lead of railroads which contracted with the Mexican government to lay tracks connecting Central Mexico and the northern mining districts to the United States. Early transnational corporations like the Phelps-Dodge Corporation reaped high profits from investments in both Arizona and Mexican copper. Phelps-Dodge owned and operated mines and smelters at Bisbee, Morenci, Douglas, and Nacozari in Sonora. Phelps-Dodge copper production rose from about 14 million pounds in 1890 to 186 million pounds in 1918.[34]

El Paso's key location as a historical way station on the Chihuahua trail, drew outside interests. El Paso became where the railroads moved American ore East and Mexican ore North. The Kansas City Consolidated Smelting and Refining Company of Argentine, Kansas, opened the El Paso Smelter in 1887 and it incorporated "at $3 million." The smelter employed 250 mostly Euro American skilled workers from other parts of the country and "unskilled" common laborers from Mexico.[35] In 1899, El Paso Smelter merged with American Smelting and Refining Company (ASARCO). Aside from the Guggenheim (Anaconda) empire, which owned mining and smelting operations in the West and Mexico, ASARCO was one of the major mining and smelting combines in the Southwest. In that year it announced profits of $3.5 million, and in 1901, ASARCO and Anaconda joined to create a smelting and mining monopoly in the Southwest. In 1902, the old El Paso plant doubled its previous output and employed 900 workers with an annual payroll of half a million dollars.[36]

Many mining towns, like railroad towns across the Southwest, were originally founded by miners and or railroad men. Anglo founders named towns after mine operators like Bisbee, Clifton, and Morenci in Arizona; likewise, they named places like Barstow and Colton (California) after railroad men. During the mid-nineteenth century Trinidad, New Mexico, and Pueblo, Colorado, originated as bicultural (Hispano and Anglo) settlements and became mining towns during the American era.[37] Many of these towns owed their existence to either mining, ranching, agriculture or the railway. According to the *Times*, "El Paso, which so largely owes its existence to railroads, derives more benefit from the payrolls of these great enterprises than from all other industries." In 1902, the monthly payroll amounted to $200,000. Moreover, the transcontinental lines that passed through the city insured El Paso's centrality to the growth of the Southwest.[38]

Railroads both stimulated and transformed agriculture in the West as they had the ranching, lumber and mining industries. The expansion of the railroad into areas that were previously inaccessible made farming available to more people, but also forced some small farmers out of business due to increased competition. The impact of the railroad increased the number of farms in the West, from approximately 2,044,000 in 1860 to 5,737,000 in 1900. According to one railroad historian, "more new farm land was brought under cultivation in the generation after the Civil War than in the entire previous history of the nation."[39]

The railroad opened up new lands for agricultural development and provided the means to transport heavy equipment for wells and mining, which propelled production. New farming techniques (dry farming, irrigation, and increased mechanization) facilitated this process and in turn created a demand for and increased dependence on the railroad for production and the market. For instance, Texas became the Cotton Kingdom of the West. Among the six major prairie farm states (Minnesota, North Dakota, Iowa, Kansas, Nebraska and

Texas) the overall railroad mileage increased from 1,000 to 42,000 miles between 1860 and the turn of the century.

Most independent farmers and cattle ranchers resented the power of the railroad and their dependency on it. This sentiment lay at the foundation of the grange and populist movements in the Central Plains states during the late nineteenth century. During the late nineteenth century, some Hispano farmers in New Mexico and itinerant traqueros joined these movements to collectively combat the power of the railroads. According to one historian, the railroad "systematically destroyed what remained of the open range." The expansion of railroad lines accelerated the transformation from herding to ranch farming.[40]

As with mining and agriculture, the railway gave ranching a boost. According to Richard White, "Abilene, Kansas, was the first of the cattle towns created by the convergence of the longhorns, the railroads, and ticks." Railroads provided access to greater markets, efficient and less expensive transportation, and bigger herds, more jobs for cowboys, and the indirect boost to a variety of small businesses that catered to the cowtown culture. While the Texas and Pacific Railroad served Abilene, other lines both founded and connected western cowtowns with other cities. In 1873, the Missouri, Kansas and Texas (MK&T) reached northern Texas, thereby linking cattle ranches there to Kansas City. Later, the Texas and Pacific connected El Paso with Shreveport, Louisiana. In 1883, the Santa Fe and the Denver and Rio Grande opened up the Southwest.[41]

Railroads also expanded the movement of livestock from Mexico into the Southwest. In 1885, railroads shipped upwards of 60,000 head of Mexican cattle, horses, and sheep through El Paso. By 1898, the American consul in Ciudad Juárez reported that the "demand for Mexican cattle, has almost exhausted the supply." While livestock suppliers came from both sides of the border, 76 percent of the entire cattle imports of the United States came from Mexico and reached $1,160,312 in 1909 according to the El Paso Chamber of Commerce. And the following year 174,530 head of cattle valued at $2,460,681 passed through El Paso representing 80 percent of the total cattle imports to the country.[42]

The consumption of beef increased dramatically between 1900 and 1914. According to Gerald D. Nash, Americans consumed 75 pounds of beef per person every year. He further states that "western cattlemen supplied the nation with about one-half of its beef requirements," and railroads provided the transportation network required for the boom in beef consumption. By 1918, American stock producers exported 500 million pounds of beef. World War I accelerated the rate of freight tonnage in 1917 and created a 43 percent increase over that of 1914–1915.[43]

The Southwest contained more subsurface sources of wealth than above ground; however, the arrival of the railroad gave tremendous stimulus to a heretofore untapped surface source of wealth. Lumbering emerged as an equally important sector of the economy in the Southwest and West, specifically for new town construction and railroad ties. As with the railroad and mining industries, a few wealthy eastern industrialists made the greatest profits. One such corporation was the Weyerhauser Corporation, which purchased 900,000 acres of timberland from the Southern Pacific Railroad in 1900.[44]

During the last three decades of the nineteenth century, railroads used approximately 25 percent of the entire timber production in the United States. In 1890, according to Richard White, "the railroads needed an estimated 73 million ties to maintain old lines and build new ones." By 1927, the railroads spent more money for timber than for any other kind of material except fuel, and consumed one-fourth of the annual timber cut in the United States. But the demand for timber diminished slightly as concrete replaced lumber for construction and coal replaced wood for fuel. Railroads also began to chemically treat ties in order to slow deterioration and thereby prolong tie life and retard the depletion of forests. Ties that previously lasted only a decade, could be replaced with ties that lasted as long as 30 to 50 years.[45]

Before 1875, railroads generally did not treat crossties until they realized that they cut timber four times faster than it grew. Railroads needed to replace ties for a variety of reasons aside from damage caused by accidents: moisture rot, decay due to bacteria, "spike killing," and "shattering." Rot occurred when moisture entered the untreated core of the tie around the spike. Spike killing occurred occasionally when a track worker drove a spike into an untreated hole and bruised or damaged the wood fibers. Shattering described the damage when workers drove spikes into unbored ties.[46]

New technology brought about the construction of tie treating plants. In 1885 the Santa Fe Railroad began treating ties in Las Vegas, New Mexico. In 1897 it opened a second plant in Somerville, Texas, and built a third plant in Bellemont, Arizona. In 1908, a new plant in Albuquerque replaced the Las Vegas and Somerville plants after they burned down. Finally, in 1924, the Santa Fe opened its fourth plant in National City, California.[47] Through the late nineteenth century, according to official records, the Santa Fe Railway had not arrived at a completely effective method for treating ties.

For the AT&SF, the scientific impact of tie treatment remained unknown until 1909. According to official records:

> The number of treated ties in Santa Fe tracks was not sufficient to effect the annual tie renewals until about 1909, but from that time until the present there has been a consistent decrease in the number of ties required each year for renewals per mile of track. During the year 1909, with 12,908 miles of maintained track of all kinds, 3,775,937 ties were required for renewals. In 1925, with 18,055 miles of track, 1,936,497 ties were used for renewals, about half the number used in 1909 and with about 40 percent more miles of track.[48]

In 1909, the United States Forestry Service and the Census Bureau reported that the nation's railroad and electric car lines purchased 40,000 fewer ties than in 1907. This report is consistent with the AT&SF report. However, it should be noted that tie purchases depended largely on the health of the economy. Railroads shipped more cargo during periods of prosperity than during depression periods. Fewer shipments meant less wear and tear on railroad ties. Less wear and tear meant that railways required fewer replacement ties and built fewer feeder lines.[49]

As noted, wealthy industrialists capitalized on the booming timber industry. But the demand for railroad ties also stimulated the local economy. In Las Vegas, New Mexico, Hispano farmers supplemented their income by providing railroad ties to the AT&SF. Indeed, as examined later in this study, Hispano night riders wearing long coats and white hoods, who called themselves Las Gorras Blancas (White Caps), threatened and intimidated Hispanos and Anglos who collaborated with the railroads. Las Gorras Blancas punished both railroads and collaborators; they cut fences, torched trestles, and destroyed crossties as an act of protest against the encroachment of the railroads whom they regarded as the cause for the loss of the Hispano sheep grazing economy.[50]

As in the West, the railroads also made Mexico dependent on the banking and industrial centers of the northeastern United States. As mentioned above, the building of the first transcontinental railroad in 1869 helped to unify the western states and territories. Government leaders worried that the geographical distance of the Southwest from the eastern centers of power might facilitate an irredentist movement in the former Mexican lands. In spite of this concern, or maybe because of it, foreign capital entered Mexico primarily to finance the construction of the railroad and then later into mining and large-scale agricultural plantations.[51]

On August 3, 1881, the governor of Chihuahua, Luis Terrazas, drove a silver spike on the American financed railroad that connected Mexico to the United States by rail line. This singular event symbolized the culmination of negotiations and concessions between the Atchison, Topeka & Santa Fe Railroad and the Mexican government to build the *Ferrocarril Central de Mexico* and the *Ferrocarril de Sonora*. Aside from the Santa Fe Railroad, the Southern Pacific also entered into negotiations with Mexico in order to build lines from Eagle Pass and Laredo, through Northern Mexico and another from Southern California down to San Blas on the Pacific side.[52]

Opposition to both the importation of foreign capital and skilled Euro American workers on Mexican railroads preceded and escalated after Mexico granted railroad concessions to the United States and England. Only two generations prior, Mexico had lost its northern territories to the "colossus to the north." One Sonora newspaper editorialized:

> The Yankee is already on her [Sonora's] frontier and asks the privilege to build a road so that thousands can easily be shipped into her borders. How long before they will come with guns and blue clothes to kill, burn, sack and annex?[53]

While this warning probably did not cause much concern at the moment, the writer was not far off the mark. In 1914, President Woodrow Wilson deployed gunboats off the coast of Veracruz and occupied the city for six months; and in 1916 General "Black Jack" Pershing invaded northern Mexico in pursuit of the revolutionary leader Pancho Villa for his 1915 raid on Columbus, New Mexico.[54]

Ultimately, foreign investment in Mexico underlay the political upheaval and insurrection of the 1910 Revolution. After 30 years of political rule by Porfirio Diaz, both the Mexican middle-class and the masses of peasants and workers rose in rebellion to overthrow the Diaz government. While the bloodshed continued for eleven more years, the Revolution failed to fundamentally transform power relations and approximately one million Mexicans migrated to the United States between 1910 and 1930 to escape the violence, hunger, and political uncertainty in Mexico. Many of these immigrants formerly held jobs as railroad shop and track workers, skilled miners, farmers, and factory workers in Mexico, and sought similar jobs in the United States. The Mexican Revolution marked the first major social upheaval of its kind in the twentieth century and is certainly understood as an important rupture in Mexican history.[55]

The social economic transformation of the Southwest is inherently tied to the processes of industrialization and modernization in both the United States and Mexico. The main rail lines built in Mexico, and financed by North American and British companies, established the foundation for Mexico's economic growth and made it dependent on the United States.[56] The prior and subsequent relations between Mexico and the United States are marked by this fundamental economic transition. The railroad directly linked work relations, family, migration patterns, trade, politics, and national development.

Between 1880 and 1900, the railroads relied on a variety of workers: Mexicans, Indians, Chinese, Japanese, native-born whites, and Europeans. And railroads regularly employed ethnic work gangs simultaneously on different sections, jobs, and operations. Railroads kept workers divided along racial/ethnic lines in part because of the linguistic and cultural differences among them, and partly because many workers themselves refused to work alongside different ethnic groups. But this racial division of labor also served to keep workers pitted against one another. In the event of a strike, railroads often replaced strikers with scabs from different ethnic groups. Perhaps more than any other reason, the high temperatures and hostile climate of the Southwestern deserts in California, Arizona, New Mexico, and Texas deterred many white men from track work.[57]

W. Thomas White has also noted this change in the makeup of the track forces in the Pacific Northwest. He argues:

> Between the completion of the first transcontinental link in 1883 (1869) and the early days of the New Deal, the Union

> Pacific, Great Northern, Northern Pacific, and Milwaukee roads all recruited laborers from widely divergent origins. Native whites and northern and western Europeans dominated the skilled and semi-skilled trades. Following the anti-Chinese campaigns and the passage of a series of laws restricting Chinese immigration in the 1880's, they [European immigrants] also dominated the unskilled section 5 end construction sectors until the mid-1890's.[58]

This transformation of labor in the Pacific Northwest parallels that in the Southwest, except that railroads recruited Mexicans into the unskilled and construction sectors earlier and more rapidly. However, railroads recruited Mexicans into the Pacific Northwest during and after World War II.[59]

The racial/ethnic segmentation of workers created a classic form of occupational and wage stratification along racial/ethnic lines. For instance, in many cases, foremen assigned Indian and Mexican workers to do grading work exclusively and paid Irish workers slightly more to lay track. Likewise, railroads preferred to employ native-born Euro American men as foremen, and bilingual Mexicans as the "straw boss."

In the United States, railroad companies hired Mexican, Chinese, Japanese, and recently arrived European immigrants to perform track work. In Mexico, railroad companies often lost their own track labor to jobs north of the border, and as a result imported thousands of Chinese and Japanese to build the Mexican lines, many of whom later followed Mexicans over to the United States.[60]

In the late 1870s and the early 1880s anti-Chinese violence signaled an important change in track labor employment in the West. The combined impact of anti-Chinese violence and cessation of the legal flow of Chinese into the West made Mexicans the next logical source of labor from the perspective of the railroads. Indeed, even as railroads and mines continued to employ Chinese (those already in the country), anti-Chinese hysteria manifested itself in riots, which literally drove Chinese workers out of towns and off work-sites.

Western journalists routinely reported on railroad construction, and in the 1870s Chinese workers appeared regularly in these reports as did the tons of equipment and supplies used by the lines. The following quotation provides a glimpse of the progress and way in which journalists understood Chinese workers:

> On Sunday came four car loads of supplies and 22 horses. On Monday, 23 horses, a car load of provisions, two car loads of turntable material, one car load of water tank material, and a lot of employees and Chinamen. On Tuesday came a water wagon for supply of outlying camps, 25 carts, 25 scrapers, a car load of tools, insulators and brackets, 67 horses, a lot of employees and two car loads of Chinamen. On Wednesday came 249 Chinamen in six cars, and three car loads of rails. On Thursday came eight cars with over 300 Chinamen, and yesterday 91 more came in.[61]

The subtle distinction between "employees" and "Chinamen" underscores the racial division of labor and dehumanization of Chinese workers as non-employees. We might also safely assume that the "employees" arrived under slightly more comfortable conditions than the Chinese workers. Indeed, railroads commonly transported "unskilled" track workers in cattle cars or boxcars regardless of their race or ethnicity. In 1882, a journalist for the El Paso *Star* provided an account of track laying for the Southern Pacific 196 miles east of the city. The report stated that there "are about three thousand men employed grading and track laying and ten carloads of Chinamen from California passed here last night to join the grading party."[62]

Like stories in Western lore, anti-Chinese sentiment took on violent proportions in isolated regions. In southern Arizona, between Yuma and Crittenden, a variety of workers, mostly Mexican but also Russian, Irish, English, Italian, and Scandinavian, took the law into their own hands in order to stop the Southern Pacific's employment of Chinese workers. In 1882, the Tombstone *Daily Epitaph* reported that

> Two attempts were made to introduce Chinese labor on the grade, and each time the celestials were driven off, and on the last occasion Mr. Scott, the Superintendent of construction, was strung to a telegraph pole by the neck until he promised to abandon the idea of working Chinamen.[63]

A local saloon keeper took advantage of the workers' isolation and monopolized the sale of "mescal and wild tarantula juice whisky." Drunkenness contributed to an already explosive level of hostility and violence. And while Mexican workers took part in the anti-Chinese violence, Euro American workers later attacked the Mexican workers when the traqueros refused the white men's demand to "surrender" their wives and daughters to them for their pleasure.[64]

In another incident near Crittenden, European immigrant workers dropped their tools and stopped a train of one hundred Chinese workers on their way to another camp site. According to an observer, the men led by a captain:

> commanded the teams to stop[,] and [commanded] the Chinamen to return from whence they came. They did not heed the warning when shooting commenced, and one could see queues and China shoes flying towards the hills. Two Chinamen were found wounded. An attempt was made to cut off one of the Chinamen's queues, but it was compromised by the Chinaman's paying a dollar.[65]

By the 1890s, the Southern Pacific had slowly replaced Chinese track crews with Mexicans on the West coast lines. Soon virtually all the railroad camps consisted of Mexican immigrant workers. Railroads occasionally employed Mexican cooks to appease the traqueros appetites as a means to retain them on the job.[66]

Indeed, throughout the late nineteenth and early twentieth centuries, the transformation of the ethnic work force in track work, mining, and agriculture continued. For instance, in August of 1883, Manuel Preciado organized a Mexican grading gang for the Santa Fe railroad connection between Colton and San Bernardino, California. Euro American laborers also worked on this grade, but the roadmaster maintained separate camps. In 1885, Mexican grading and track laying gangs also worked alongside Chinese and white workers in constructing the Santa Fe railroad through the Cajon Pass north of San Bernardino to Barstow. According to Santa Fe records, Chinese workers constituted the bulk of construction gangs on the Cajon Pass construction. And as will be shown later in this chapter, the Santa Fe shipped in additional Mexican workers directly from El Paso as they needed them.[67] The relatively low cost and abundance of Mexican labor allowed the Santa Fe railroad to replace its more expensive Euro American labor with Mexican labor.

This replacement of white and Chinese workers with Mexican workers played itself out repeatedly throughout the Southwest. For instance in 1898, the *Arizona Silver Belt* reported that railroad "construction crews are composed chiefly of Mexicans and Indians, with a few whites. It is difficult to get white men to work, the wages being only $1.50 per day, and board $5 per week with some minor charges, which reduce a man's net earnings."[68] In 1902 the *Los Angeles Socialist* paper editorialized that

> White men cannot compete with the Mexicans and provide [for] themselves with families at the same time. In the competitive race the Saxon has given up all hope of posterity, [*sic*] and it is well, for the race had better die out rather than be turned into beasts of burden.[69]

Since 1903, a Santa Fe official reported that Mexicans "have been used exclusively in the track department of the coast lines."[70] In 1908, a Department of Labor report noted that "In some occupations Mexicans are rapidly displacing Japanese, Greeks, and even Italians."[71] According to Mary Roberts Coolidge, "The Cholos . . . were brought into Southern California because neither white nor Chinese labor could be obtained for railway and construction work."[72] L.L. Waters, Santa Fe railroad historian put it tersely, "Paddy [e.g. the Irishmen] was replaced on the section by Mexicans."[73] In 1910, the Dillingham Commission investigated labor conditions among immigrants in the West and Southwest and found that the Santa Fe Railway Company "is at present one of the largest employers of Mexican laborers."[74] And according to the federal census of 1920, Mexican workers entered the United States at the rate of about 100,000 per year. Likewise, in 1925, *The Los Angeles Times* reported that Mexican workers replaced Japanese who moved to the cities.[75] This racial/ethnic transformation of track labor led to what political economist Harry Braverman calls the degradation of labor, and will be discussed further in the following chapter.[76]

As railroads expanded their employment of Mexican workers, white workers sought better pay and preferred jobs in the railroad shops or in other employment sectors. In the mid-twenties, a study concerning the impact of Mexican workers on Texas summed up the reasons why white workers despised track work. It noted:

> White laborers feel themselves superior, say to negros or Mexicans and will be more or less dissatisfied to remain in a

> [racially] mixed group; they will, therefore seek other employment. Once negros or Mexicans are employed on section gangs, it will be a matter of time, only, until the white men leave them.[77]

The *Railway Age Gazette*, a respected trade association journal, acknowledged the Southwest to be the most difficult region of the country to recruit quality workers. Another observer captured the central reason why white workers avoided track work, especially in the Southwest, "They [Mexicans] are doing work that the average American would not and cannot think of doing. Last week at Bagdad in the desert the heat killed three Mexicans and prostrated a score more. The temperature was up to 120 degrees." The implicit identification of the "average American" as a white man is obvious and helps to explain the ethnic transformation of the track labor force.[78]

During the economic downturns of the 1920s, Mexicans numerically dominated the field of track labor and some politicians used the racist premise that Mexicans took jobs away from white workers to argue for restrictive legislation. Both the railroad's demand for "cheap labor" and the racial hierarchy that stemmed from the employment of Mexicans almost exclusively as track workers redefined this job category as "Mexican work," which the majority of white workers avoided at all costs. Alexander Saxton offers a colorful description of both the social hierarchy of the railroad business and the gradual transformation of the racial/ethnic makeup of the track labor force:

> As if in a daguerreotype from the first decade after the Civil War, the characteristic picture appears: the face of the engineer, always white, capped and goggled at his cab window, one gloved hand raised in casual greeting to the gandy dancers [track workers] along the trackside—formerly Chinese or Irish, now generally Negro and Mexican—as they stand back to let the express roll through.[79]

Railroads employed Black track workers for the most part in the South and the Central Plains, and in places like Galveston and New Orleans while Mexicans and Blacks often worked in the same track gangs. They also competed with one another for track work.[80]

Railroad construction and maintenance-of-way was by nature temporary, seasonal, and low paying. According to Robert Athearn, during the early days of the railroad in the East, white native-born workers regarded track workers as the nobility of the road.[81] But with the shift of European immigration from western Europe to eastern and southern Europe during the late nineteenth century this image slowly changed. By the 1880s Eastern European immigrants, Mexicans, Indians and Blacks employed on the railroad completed the degradation of track work. This pattern of racial ethnic stratification was firmly entrenched by the time the first big waves of Mexican immigrants arrived and found employment on American railroads. Moreover, the breadth of the Mexican diaspora in the railroad industry made track work unattractive to native-born workers, including many Hispanos.

In 1900, America's railroads employed 1,018,000 people, and with some fluctuation, the number increased to 1,786,000 in 1925. This number represented 6 percent of the country's non-agricultural labor force.[82] Traqueros like Jesús Ramírez represented the most invisible sector of the railroad industry, despite the fact that between 1880 and 1930, Mexican track workers constituted almost two-thirds of the track labor forces in the Southwest, Central Plains, and Midwest. Jesús Ramírez possessed an innate sense of the significance that his life and work represented. Thus, like Ramírez, the present study seeks to fill an important gap in the history of the West and the Southwest in American working-class history.

chapter 2

Labor Recruitment

"THERE WAS MUCH WORK TO DO FOR THE RAILROAD," FORMER traquero Jesús Ramírez recalled.[1] Ramírez was born in Silao, Guanajuato, in 1900, and left at the age of fifteen with his father to lay tracks in Kansas, working ten hours a day at ten cents an hour. Because of the lack of work in Mexico, and the unsettled conditions resulting from the Revolution, hundreds of thousands of Mexican men and women decided to leave Mexico, at least temporarily in order to find work, peace, and to raise families in the United States. Ramírez eventually remained with the Atchison, Topeka & Santa Fe Railroad in Emporia, Kansas, until he retired. When Ramírez arrived in 1916, annual Mexican immigration hovered around 18,000. By 1920 that figure had jumped to 54,000.[2]

This chapter examines the origins, growth, and diaspora of Mexican railroad workers in the United States, especially on the Atchison, Topeka & Santa Fe. Within the constraints presented by the growth of the railroad in the Southwest, Mexican track workers and their families made important decisions about the conditions of their employment, where they worked, where they lived, and how they determined the quality of their daily lives within the confines of a burgeoning extractive economy.

Mexican immigration to the United States during this fifty-year period was a complex web of relationships between Mexican workers and American railroads. On the one hand, the economic transformation of the Southwest from a pre-capitalist economy to a full-blown industrial one created an insatiable demand for cheap labor. On the other hand, socioeconomic and political upheaval in

Mexico gave rise to the displacement of tens of thousands of Mexican men and women from cities, towns, villages, haciendas and *ejidos*. Approximately one-eighth to one-tenth of Mexico's population, or one million people, shifted north of the border during the period 1900 to 1930.

Railroad construction into the West transformed the region's economy and created a seemingly insatiable demand for labor. Combined with the growth of the mining and agricultural industries, the railroads found themselves in stiff competition with other industries for the same pool of laborers; however, railroads also had a vested interest in supplying other industries with Mexican labor. The railroad's primary purpose, after all, involved the shipment of raw materials and goods out of and into the West and Southwest. And they profited handsomely. This demand for track labor accounted for the majority of "common labor" jobs in the West.[3]

In the 1870s and 1880s, railroads, mining, and agriculture made use of the available resident pool of labor in the Southwest: Hispanos, Indians and European immigrants. During this period, Mexican immigrant workers occasionally sought work on the north side of the border. As the demand for labor expanded, American railroads actively recruited mexicanos to work north of the border. This demand for labor, on the one hand, and the need for work, on the other, created a circular migration into and out of the United States, but authorities paid little attention to this phenomenon until well into the 1900s.[4] Indeed, it appears that the migration trail of most Mexican railroad workers to the United States originated in the densely populated central region of Mexico, and from there to the industrialized northern states of Chihuahua and Sonora from where transnational migration continued. Mexican immigrant labor joined, replaced, and eventually outnumbered domestic Hispanos and all other immigrant track workers (including Black workers) in the Midwest and Central Plains.[5]

The demand for track workers multiplied with every new mile of track laid, and the demand grew exponentially once completed. Branch lines, interurban lines, double-tracking, and maintenance kept tens of thousands of traqueros employed almost year-round. Mexicans sought jobs that paid the best in order to improve their living conditions. And like Hispanos before them, they sought railroad work because it allowed for seasonal transnational migration that allowed them to earn a modest income and return to their homes. Of course, these same features frustrated others who sought long-term and more secure employment. Track work in Mexico paid comparatively less, roughly twenty-five cents a day, while the same work paid one dollar a day north of the border throughout the period from 1880 to 1910. This wage must also be understood in comparison to the wages the Union Pacific paid Irish immigrant workers on the first transcontinental railroad: three dollars per day.[6]

Although work on the American side paid better, housing and food also cost more in the Southwest than in Mexico and Mexican workers quickly came to terms with this reality. Thus, railroads preferred to hire Mexicans who had no previous experience doing track work. "Greenhorns" typically did not protest working conditions, nor did they often complain about low pay or overpriced food and supplies. They also worked harder to impress the foremen, whereas more experienced traqueros protested virtually everything and cleverly figured out how to slow down the work process and avoid overexertion. The alleged "docility" that railroads praised Mexicans for must be viewed in this context. Subaltern resistance will be discussed in Chapter 4; however, the tendency for traqueros to spontaneously walk off the job (both individually and in groups) also contributed to the demand for labor.

Unfortunately, historians have not yet determined the exact numbers of traqueros who walked off the job, but the rate was sufficiently great that employment agencies and railroads later attempted to make track work more attractive to Mexicans. Many if not most of the traqueros who walked off the job found work elsewhere almost immediately in industrialized areas of the Southwest, Central Plains, and Midwest. Some found better paying and more lucrative employment in meat-packing plants, foundries, steel manufacturing, mining and smelting plants, and even the auto industry.[7] Others found employment on the farms and canneries. Ultimately, the oppressive conditions on the railroads provided much of the incentive for traqueros to find work elsewhere.

Railroads functioned as a conduit for the movement of Mexican workers from one occupational sector to another. Traqueros, like thousands of other common laborers during this period, "tramped" (rode the rails) around the country as "hoboes." Some even joined the famous Industrial Workers of the World (IWW), also known as the Wobblies. Wobbly lore has it that some trainmen honored the little red IWW membership cards, which saved "bindlestiffs" (hoboes) and "gandy dancers" (track workers) from being tossed out of a rolling freight car.[8]

Indigenous groups of the American Southwest traveled and traded extensively across this region before and during European contact. However, it is less acknowledged that many of these trade routes were the original paths upon which American and Mexican railroads laid track.[9] The movement of people and goods originally begun by indigenous groups was thus merely replaced over time by the Spanish, Mexicans and Anglos.[10]

These trading patterns changed substantially with the industrialization of both northern Mexico and the Southwest. Notwithstanding the national boundary between the United States and Mexico established in 1848, the uneven but parallel development of northern Mexico and the American Southwest is best understood as a single socioeconomic region as opposed to two separate and unrelated national developments. As shown in the previous chapter, foreign investment provided the technological and financial means for Mexico's economic growth. The year 1880 represents an equally important watershed because the history of the Southwest can usefully be divided between the pre-railroad and railroad era in terms of economic development (mining, ranching and agriculture), proletarianization, labor recruitment/migration, and community formation.

Prior to 1869, the Central Pacific briefly considered importing "thousands of peons" from Mexico but Euro American opposition in California ended that possibility. Thus American-born Spanish-speaking workers from New Mexico—the Hispanos—constituted the first "Mexican" railroad workers in the Southwest as early as 1871. Heretofore, scholars have argued that railroads did not employ Mexicans until the 1880s and early 1900s. In August of 1874, Santa Fe Railroad payroll ledgers list numerous Spanish-surnamed "laborers" in Colorado.[11] During this time the Southern Pacific also entered the Southwest through East Texas and employed some of the earliest Mexican-American railroad-workers in that state: the *Tejanos*. However, official Southern Pacific documents for the same period are not available, and thus it would appear that the Santa Fe Railroad employed Mexican-American workers before all other railroads. In both cases, the Santa Fe and the Southern Pacific Railroads employed Mexican-American workers at approximately the same time.[12]

The identification of all Spanish-surnamed workers as Mexican as opposed to Indian or Hispano may be imprecise. It cannot be stated absolutely certain that workers of Indian, *mestizo*, and Hispano background self-identified under every circumstance as Mexican, but they nonetheless represent the basic elements of Mexican identity. Although it is not widely acknowledged, some Southwestern Indians also found work on the western railroads as early as 1878.[13] Nonetheless, according to Samuel Bryan, the Mexican immigrant population in the Southwest increased from 66,312 to 99,969 between 1880 and 1900.[14]

Spanish-surnamed Indians are literally impossible to distinguish from Mexicans through the documents. Many Pueblo Indians possessed Spanish surnames as a result of Spanish colonization and Mexican settlement. Railroad payroll ledgers and newspaper accounts almost never differentiated between Mexicans and Spanish-surnamed Indians.[15] Depending on the context and circumstance, these groups commonly self-identified as Sioux, Navaho, Apache, Pueblo, Yaqui, Español, mexicano, or some combination thereof. Further complicating the issue of ethnic identity are the numerous Yaqui, Maya, and Zapotec Indians, among others, from Mexico who later found work on American railways. An official report in 1910 noted the "efficiency" of Yaqui Indians, and yet, according to the report, Yaqui's did not "mix" well with Mexicans.[16]

In the early years of track construction through New Mexico, Santa Fe Railroad officials made verbal agreements with the Laguna Pueblo Indians to hire all those Lagunas who wanted to work on the railroad indefinitely in exchange for the right to build track through traditional Laguna lands.[17] Documentation on this agreement is scarce, but the Santa Fe Railroad appears to have respected the agreement and frequently hired Pueblo Indians to the exclusion of Mexicans. While Indian track labor and shop laborers increased only slightly, the Santa Fe Railroad employed them as track workers consistently throughout the nineteenth and early twentieth centuries. From the 1880s onward, Mexicans outnumbered Indians and other groups at any one time on the tracks, and the gap between the numbers of Mexicans and all other groups subsequently widened.[18]

As with Mexican workers, the Santa Fe and other railroads reported inconsistent and contradictory information regarding the capability of Indian workers compared to other racial/ethnic groups. Some adapted well, and others did not. In either case, the frequency by which Indian workers sought and maintained regular employment defies the old biased assumption that the Indian of the "past" had no place in technological America.[19] Still, more research on Indian railroad workers is needed to understand how and to what degree routinized work processes affected the various Southwest Indian cultures.[20]

In the early 1870s the Santa Fe organized track crews in Kansas westward through eastern Colorado and then south to the northern edge of New Mexico and as such included a mix of European ethnic workers imported from the Midwest. Laying tracks in the Rio Grande valley of New Mexico widened industrial development, including mining, commercial agriculture, lumbering, and, of course, track work itself required a labor supply.[21]

According to Sarah Deutsch, railroad work allowed some Hispano farmers to make a success of their homesteads in a rapidly expanding economy. She argues that Hispano villagers took advantage of the episodic employment on the railroad to expand into territory governed by "an Anglo system where neither agricultural land nor pasturage was free."[22] While some Hispanos may have welcomed the opportunity to work on the railroads, it is doubtful that the majority welcomed or profited as much as their Anglo counterparts from such employment. Damacio and Laretta López provide a good example of the strategy for survival outlined by Deutsch. In 1871, López, along with eleven other Hispanos, filed for homestead land in Red Rock, Colorado. Like other Hispano farmers, the men planted and harvested their crops communally and shared the responsibility of labor. Aside from his crops, López owned about 15–20 head of cattle, but could not survive on his holdings alone. In 1876, he and Laretta moved to West Las Animas, Colorado, and worked on the grade with several hundred other Hispanos at the southern railhead of the Santa Fe. Track work allowed López to accumulate enough savings to return to his homestead in Red Rock with more cattle and open a small store.[23] This pattern of episodic employment for Hispanos repeated itself as they built the Santa Fe southward through New Mexico.

The Santa Fe recruited locals through a variety of methods: newspapers, word-of-mouth, and handbills. It seems reasonable to assume that Hispanos also shared information through their social networks in villages, taverns, churches, and with friends, relatives and neighbors. No doubt they also shared information about particularly good or bad track sections, foremen, food, etc., in bunkcar conversation. This practice was so widespread that by the second decade of the twentieth century, Euro American foremen and roadmasters discussed the issue of a Mexican "clearing house" openly in the pages of trade magazines. For instance, one general foreman in 1914 had this to say:

> Most Mexicans cannot read or write. Consequently they pass a great deal of their time, when in bunk cars, talking about the different places at which they have worked. Any section or extra gang boss who is either very good or very hard on his men is known all over the system through this "clearing house."[24]

Despite the erroneous generalization regarding Mexican or Hispano literacy, this quotation provides an important glimpse of at least one mechanism by which Hispano workers determined for themselves where they might find employment.

Hispanos also found work in various sectors of the railroad-stimulated economy. For instance, saw mills in southern Colorado employed Hispanos producing railroad ties and lumber for trestle and building construction. Some also worked as freighters (*arrieros*) and ranch hands (*vaqueros*).[25] According to Richard Nostrand: "Most of these Hispanos were men who were away from their villages temporarily or seasonally, and all worked for Anglos."[26] Like Deutsch, Nostrand found that by the 1890s Hispanos had developed a vibrant network between villages and employment sectors of the Rio Arriba region and southern Colorado. This cyclical migration extended down to the border region and included Mexicans and Yaqui Indians. Moreover, this form of migration both preceded and continued through the major waves of the turn of the century and early 1900s.

Not all Hispanos worked for Euro Americans; some worked for themselves. The railroad also created a demand for individual "tie hacks" and small Hispano-owned saw mills. Tie hacks used only a sharp axe to fell trees and cut ties. Depending on the conditions—tree size, type of wood, and the condition of the axe—an exceptional tie hack produced about 60 ties per day. Moreover, hauling ties from the countryside to the railroad yards also required tremendous upper body strength and draft animals that needed to be fed and tended to. Some local lines also used Mexican tie hacks to clear brush and trees from the right-of-way.[27]

In the 1880s, railroads paid five cents per tie; thus a tie hack could theoretically make three dollars a day. However, tie hacking could only be done seasonally and therefore limited year-round self-employment. On the other hand, this seasonal work also allowed Hispanos to tend to their crops or find other more remunerable employment.[28] Of course, the growth of the railroad into Colorado and New Mexico gave rise to other less cooperative responses, especially when the railroad and Anglos abrogated traditional water and land rights. As noted in a later chapter, *Las Gorras Blancas*, otherwise known as the "white caps," perhaps stand out as the best example of organized violent resistance to the railroads (see Chapter 4).

Prior to the massive influx of Mexican immigrant workers after 1900, Hispanos wrote *corridos* (folk ballads) about track work. Mexican in form, and sung in Spanish, corridos reflected the geographical and social contact with American institutions and people. In this sense, Mario García correctly points out that the immigrant corridos comment on what he terms the "immigrant experience."[29] However, the corrido below recorded by New Mexico anthropologist Charles Lummis reveals a non-immigrant experience with the railroad. As New Mexico natives, the railroad came to them. Hispanos through their folksong (the corrido) commented on jobs as well as the opportunity to sell crafts, homemade food, and other items:

The Railroad

The railroad is coming this way—
Let us go look at it near.
When we shall see it appear,
Ah, what a joy it will be.

Chorus: And when the tourists shall throng,
"Good morning!" I will repeat;
"Come in! Come in!" I'll entreat,
"Come ye and list to my song!"
Up from the town [El Paso] on the line
come running the Americanos,
Earning us everyone money—
Money for all us paisanos.—*Chorus*.

Were you at the camp where they're giving
Work, and the laborers bringing—
That's just the way I live singing

only to earn me a living.—*Chorus.*

Neither sun, neither wind shall nor could
Make me turn back till I've done;
Mile after mile I will run
That I may win us some good.—*Chorus.*

The engine is going to start.
Lively! Be all of you ready!
Come, pull your money out—steady,
For now we are going to start!—*Chorus.*[30]

This corrido reflects an entrepreneurial attitude that not all Hispanos shared about the presence and influence of the railroad. Optimism notwithstanding, this ballad also reveals both the new status of New Mexico as a stop-off point for American tourists traveling through the Southwest, and the new dependency of Hispanos on the railroad for both tourism and jobs.

Hispano tie hacks, freighters, cowboys and trackmen began a process of work and migration in the middle valley of New Mexico and across the Rocky Mountain region that persisted throughout the late nineteenth and early twentieth centuries. Indeed, Mexican immigrant workers found jobs that American-born Mexicans and borderland migrants had been performing for the previous two decades. The ebb and flow of this cyclical migration continued unevenly throughout the early twentieth century.[31]

The railroad created a demand for labor and thereby compelled Hispanos to find temporary wage work on the railroad or related industries. This demand for labor soon depleted the local supply of Hispano workers, and railroads looked to Mexico for track labor. The convergence of American and Mexican railways in El Paso in 1881 marked the beginning of a dramatic rise in the employment of Mexicans on American railroads.

Like other immigrants to the United States, Mexican immigrants made extensive use of their kin, extended kin, village and community ties to help them determine their points of destination and occupations once they arrived. Chain migration also allowed immigrants to maintain direct ties with family and friends in Mexico, returning to the United States often with a friend or compadre. This process is consistent with the insights provided by Herbert Gutman's critique of the "breakdown model" of immigrant groups. Gutman correctly emphasized the adaptive capacities of immigrant groups rather than their cultural breakdown and assimilation, as popularized by historians of immigration to the United States.[32]

In an attempt to accelerate the assimilation of Mexican immigrant workers into American culture, the Santa Fe Railroad altered its employment policy. Under the leadership of President E. P. Ripley, the Santa Fe attempted to implement some of the principles of scientific management in order to increase efficiency and loyalty. In this way, the Santa Fe Railroad provided an institutional framework for chain migration.[33]

Some of the basic changes involved the retention of their best and most loyal track workers. Toward this end, the Santa Fe and other companies encouraged their employees to bring their wives and families with them to live on company property rent free. Housing usually consisted of an empty boxcar fitted with a wood stove. Some companies provided old railroad ties to be used for fuel, and water was usually pumped into a communal faucet. Outdoor toilets were also provided and the company typically paid for sanitary service.

In exchange for the above amenities, the companies required loyal service from their Mexican employees. By encouraging Mexican women and families to immigrate to the United States, Ripley wanted to recreate and reinforce Mexican patriarchal relations in the home in order, he thought, to thwart the influence of radical labor unions like the Industrial Workers of the World as well as to stunt other forms of independent labor organization. Ripley's views paralleled those of other industries in the West who regarded the absence of women as the root cause of social ills, radicalism, unionism, and homosexuality among their male workers. Thus Ripley and others interested in scientific management thought that Mexican women were needed to civilize the Mexican work force.[34]

It appears that Ripley successfully created a fairly docile group of Mexican employees who felt that the Santa Fe Railroad was genuinely interested in their welfare. However, this policy had unexpected and unintended consequences. This paternalistic policy actually increased the rate of both family and individual immigration to the United States and contributed to the numbers of Mexican immigrants settling in Santa Fe Railroad housing.[35]

Individual traqueros and their families also made key decisions about their work lives, albeit within the framework of industrial capitalism. Scholars have documented that Mexicans in the United States have not been passive objects in history, including those who have come as immigrants.[36] Mexican men and women used their social networks of relatives, friends, compadres, and neighbors, to find jobs on specific railroads, divisions, sections, and work gangs. Thus, like other immigrant groups, traquero families developed migratory chain and occupational linkages. For example, Domingo López of Topeka, Kansas, recalls coming to the United States with his father as a teenager during the Mexican Revolution because his uncle found lucrative work on the Santa Fe. The senior López eventually sent for his wife in Silao, Guanajuato, and lived in one Topeka boxcar barrio they called *La Yarda*.[37] In some Mexican villages, almost everyone had relatives in a sister city in the United States. Silao, Guanajuato, and Topeka, Kansas, is one such example. Another example is that of Pascual Aparicio who crossed the border at El Paso in 1919 with his parents and one-year-old brother. His father Mercedes Aparicio was a miner at La Bufa in Zacatecas and he and his wife Anita worked their way on the tracks to Colton, California, in order to eventually obtain work at the Colton Portland Cement mine. When they arrived they stayed with their cousins in South Colton and Mercedes Aparicio went to work at "la planta de cemento" (the cement plant) immediately. Like other traquero families, they built their first house out of scrap materials from the Southern Pacific and Santa Fe yards.[38] And like many other traquero families, they facilitated the continued migration of family and friends from Zacatecas.

Some railroads also routinely provided pencils, official stationery, and postage stamps to Mexican workers in order to correspond with family members and friends left behind south of the border. Foremen instructed traqueros to invite family and friends to work on specific lines, divisions, or sections. These workers wrote of jobs, housing, and free transportation from the border to a specific destination. In this way, railroads facilitated chain migration. A former traquero from Kansas recalled:

> Well, they were all good men, they gave us paper and pens to write home. Stamps, to write home to say that I found work. We had work, for others to come ready to work. And now, no, there is no work, or anything like that. Now there are no *reenganches* [labor recruiters] or anything. This has all ended.[39]

This quotation also reveals how railroads avoided violation of federal laws prohibiting labor recruitment in foreign countries by American companies, specifically the Contract Labor Law of 1885. They did this by converting their own workers into informal labor contractors.

In Colton, California, traqueros and other Mexican workers made regular use of the postal services in order to send money and to communicate with family and friends in Mexico. Mexicans packed post offices on Sundays to the disdain of impatient Euro American patrons who disliked waiting in line with Spanish-speaking Mexicans. They demanded and successfully required that the post office provide a separate window for Mexicans in the back of the building.[40]

Zaragosa Vargas, in his study of Mexican workers in the Midwest, has shown that some highly educated and ambitious Mexican track workers quickly abandoned railroad work once they learned about better paying and more secure jobs in the foundries, meat-packing plants, and auto industries in the Midwest. Those who had enough money and factory skills headed directly for Detroit and avoided track work altogether.[41]

By the 1920s, Mexican immigrants in places like Detroit told their families and friends in Mexico to avoid track work altogether. To these men and women, track work was for "greenhorns" who had few networks to assist them to find better jobs. Rather than settle for track work, they sent money to their relatives for rail passage through to Detroit in order to have their friends, compadres, and relatives work alongside them in the foundries, steel plants, and other manufacturing industries.[42]

As with other immigrant communities, traquero communities served as springboards for both acculturation and for further immigration of friends, family, and neighbors from Mexico. Many Mexican villages sent family and friends to specific barrios in the United States to work on specific divisions and track gangs. Others who arrived merely stayed in boxcar communities temporarily while they found jobs in other employment sectors.

Of course traqueros found various ways to resist unscrupulous employers and labor contractors. Even if Mexican track workers had no family network to assist them, they rarely ventured into track work alone. In most cases traqueros traveled with friends, compadres, or relatives while working on the track. And occasionally, women accompanied their husbands on the track along with children. It was not uncommon for relatives to travel together as was the case with Francisco and Nellie Quinn.

In 1918, Francisco and Nellie, with their infant son Anthony, the future film star, rented a small house with their cousins and Francisco's mother in El Paso. The men worked as day-laborers until one day Francisco returned home and discussed plans for work on an *enganche* (work gang) for the Southern Pacific Railroad. Nellie recalled:

> Down at the freight yards, there were hundreds of people yelling and screaming to get on the trains. Glamis [California] sounded like paradise to them. Even though we were all being pushed into dirty cattle-cars, everyone was begging to go. Francisco and Glafiro had these papers and we were allowed on. There were people there who were willing to buy those pieces of paper.[43]

She continued, "Half the trainload was young children, screaming and crying. You can imagine the smells and the filth on that train... I don't remember how many days and nights we traveled. The trip seemed to take forever."[44] When they arrived at Glamis, California, a foreman met them and pointed them toward the only buildings in town, old boxcars on the siding. Francisco ran to choose one of the boxcars for his family. Nellie recalled, "Inside the boxcar it was like an oven. Outside it was enough, but inside was hotter than any hell I can imagine."[45]

Of course not all traqueros had access to family networks who could guide them to specific jobs and opportunities. Some relied on less sophisticated methods to determine occupational choices. During the 1920s, for example, Chencho Alfaro of Topeka recruited Mexican workers for the Santa Fe Railroad in South Texas. When asked why the workers chose the Santa Fe, he said that the white cross emblazoned on the side of the boxcars struck a religious chord with many traqueros. They felt that the railroad called "Santa Fe," meaning holy faith, might offer better treatment and protection from the harsh realities of daily life in the United States.[46]

Increasingly throughout the last two decades of the nineteenth century, the patterns of railroad recruitment and Mexican immigration into the Southwest and beyond became consistent. Railroads, mining, and agricultural interests began by sending agents into Mexico in order to recruit Mexican workers at farms, railroads, haciendas, cities, mining districts and villages with promises of great wealth and free transportation. Indeed, this early recruitment effort catalyzed a flow of Mexican migration to and from of the United States that continued throughout the twentieth century. And, despite restrictive legislation like the 1885 Contract Labor Law and the 1917 and 1921 Immigration Laws, the relationship between American industrialization and Mexican labor became inextricably tied and co-dependent.[47]

Railroads viewed Mexicans, especially "Mexican peons," as the best source of inexpensive labor for railroad construction, maintenance-of-way, section work, and extra gangs.[48] First, Mexicans filled the demand for labor generated by the anti-Chinese movement. Unlike Chinese workers, Euro Americans expected Mexicans to return to Mexico when the track work season ended. Secondly, roadmasters immediately prized Mexican workers for their alleged docility and willingness to work cheaply. Finally, while Mexican immigrant workers often abandoned railroad work for other employment, especially in agriculture, railroads benefited indirectly by shipping farm products harvested by Mexican laborers.[49]

Typically, railroad companies or their agents (called *reenganchistas*) hired Mexican laborers in major hub cities like El Paso, Laredo, San Antonio, Ft. Worth, or Los Angeles in enganches and promised them free return passage to the border if they worked continuously for six months. Most railroad foremen much preferred to hire experienced track workers instead of greenhorns. However, this preference stood against the interest of roadmasters and employment agencies who preferred workers who would not protest contractual conditions, strike, or walk off the job. Greenhorns would "stick" better than experienced trackmen, but seasoned track workers demanded better wages and fewer hours. Thus in order to retain expert traqueros, railroads continuously searched for incentives to retain their best workers. These incentives ranged from a modest bonus at the end of six months, to "free" housing, or merely free return transportation to El Paso or some other destination along the border. Ultimately, the railroad corporation's demand for cheap labor prevailed over foremen's desire for experienced track workers.

The stereotype of Mexican labor docility is more a reflection of how railroads preferred to perceive Mexican workers than an acknowledgment of the nuanced complexity of the Mexican labor force. This distortion appeared to reify the railroads' continued insistence that they be allowed to hire Mexicans along the border unencumbered by federal authorities, who frequently accused railroads of discriminating against white labor in the maintenance-of-way department.

Another stereotype about Mexican track workers stemmed from the peculiar conclusions of railroad officials and farmers. In his 1925 study, Charles H. Hufford found that Mexicans in Texas were uniquely suited to desert work. He asserted:

> They know the art of subsisting in the desert whether it be obtaining water from hidden sources, kindling a fire under adverse conditions, curing meat and preserving food, cooking with improvised implements, constructing a hut with mud and sticks and thatch, or making use of medicinal herbs to alleviate their ills—in all they are wonderfully adapted to the peculiar conditions under which they must live as laborers in the development of the Southwest.[50]

Hufford's benignly racist conclusion that Mexican workers were "wonderfully adapted" to the hot and arid conditions of desert work in the Southwest is examined more thoroughly later. It is interesting to note that Mexican culture is linked to what appears to be a natural condition at the lowest rung of the occupational ladder and therefore reinforced the concentration of Mexican labor in track work.

The patterns of Mexican labor recruitment and immigration were established in the early 1880s. At first the railroad employed Mexicans directly and sent their own agents to the border and sometimes directly into central Mexico to recruit Mexican workers for employment in the United States. Later, by 1900, different railroad companies contracted with specific employment agencies to recruit, transport, feed, and provide housing for Mexican workers. White and Mexican American labor recruiters literally stumbled over each other at the El Paso border as they competed to enlist newly arrived Mexican immigrants. Called "enganches" or labor conscripts, labor contractors provided the labor for a given price to the railroad for a specific job and were thus not directly employed by the railroads.

In exchange for the contract, the employment agency was granted the right to profit from the exploitation of Mexicans by paying these workers in scrip, compelling them to buy clothes, tools, and supplies from the company store. Mexican workers were also expected to have their meals in the dining car with the cost deducted from their pay. And since the company store exorbitantly priced its merchandise and dining car meals, the store compelled workers to buy on credit. Workers either paid off the debt, carried it to the following job site, or jumped off the train and abandoned the company to escape these oppressive conditions. This unremitting indebtedness continued unabated for the most part through the last two decades of the nineteenth century, and the rate of desertion by traqueros rose significantly.

The exact number of Mexican track workers on the main lines of the Southwest during the 1880s and 1890s seems impossible to ascertain because of incomplete payroll records and the frequency by which census enumerators neglected to count itinerant Mexican trackmen. Moreover, Mexican labor recruitment and immigration are historically linked and impossible to separate analytically. It follows that the number of Mexican immigrants passing through El Paso, and other border crossings, would suggest approximate numbers of Mexicans on the railroads (see table 2.1). However, because immigration officials did not keep accurate records of every Mexican who traversed the border it is difficult to establish the exact number of Mexicans entering the United States between 1880 and 1930.[51] According to a 1908 Department of Labor study, "Complete statistics of those who cross the frontier are not kept, but the number does not exceed 100,000 per annum, and probably is not over 60,000."[52]

TABLE 2.1
MEXICANS ADMITTED TO THE UNITED STATES, 1894-1920

1894	109	1903	528	1912	23,238
1895	116	1904	1,009	1913	11,926
1896	150	1905	3,637	1914	14,614
1897	91	1906	1,997	1915	12,340
1898	107	1907	1,406	1916	18,425
1899	161	1908	6,067	1917	17,869
1900	237	1909	16,251	1918	18,524
1901	347	1910	18,691	1919	29,818
1902	709	1911	19,889	1920	52,361

Sources: U.S. Immigration reports as quoted in Charles H. Hufford, *The Social and Economic Effects of the Mexican Migration into Texas*, 33; Ricardo Romo, "Mexican Workers in the City: Los Angeles, 1915–1930" (Ph.D. dissertation, University of California, Los Angeles, 1975): 65; Mario T. García, *Desert Immigrants* (1981): 36.

In 1912, Samuel Bryan, a Stanford University economist, argued that until 1908, the figures for Mexican immigration "are so obviously incomplete as to be of little value."[53] Nonetheless, payroll records for the Santa Fe Railroad provide us with a sense of the dramatic increase of Mexican track laborers in the railroad town of Argentine, Kansas, from 1905 to 1927 (see Table 2.2).[54]

TABLE 2.2
MEXICAN TRACK LABORERS IN ARGENTINE (KANSAS CITY DIVISION), 1905–1927 (JUNE SAMPLE)

Year	Total Mexican Employees	Total Number of Employees	Percentage of Mexican Track Laborers
1905	52	803	6.51%
1910	240	435	55.2%
1915	366	429	85.3%
1920	305	349	87.4%
1927	366	400	91.4%

Source: J. R., "Outline of Study. Mexican Track Labor on the Atchison, Topeka and Santa Fe Railroad," Topeka, Kansas, AT&SF Railway, General offices, August, 1928. PTC. Also cited in Laird, "Argentine," 117.

After the 1890s, we may safely assume that the numbers of Mexican track workers on the major Western lines during the normal and peak track maintenance season hovered at approximately 50 to 70 percent of both regular section gangs (semi-permanent gangs responsible for maintaining particular sections of track) and extra gangs (seasonal gangs sent to wherever needed). Numbers for the first few decades of the twentieth century are slightly more reliable, but must also be examined critically, for the same reasons: economic depressions, poor enumeration, cultural bias, and specifically the itinerant and migratory work patterns of Mexican males and families.[55] For example, the Southern Pacific employed Mexican workers exclusively on the lines between El Paso and Los Angeles, but in 1901 listed only 376 Mexican track workers in California, or 11 percent. Meanwhile, Chinese and Japanese track workers comprised 9 and 4 percent respectively, with "whites" constituting the remaining 76 percent. The Industrial Commission of the federal government noted that the Southern Pacific often listed Mexicans and some "Asiatics" as "Americans" and therefore skewed the number of "whites" listed as track workers.[56]

In 1911, the Dillingham Commission found that among the nine western and southwestern lines the Santa Fe and Southern Pacific railways employed most of the so-called common laborers from Mexico. Most traqueros worked in the desert region along the border and fanned out across the West, Central Plains and Midwest. Out of 2,672 common laborers employed by the Santa Fe, Mexicans represented 2,599. The Southern Pacific employed 2,714 Mexicans out of 12,592 common laborers. The Commission also pointed out that "The number of Mexican section hands employed in the Western States has more than doubled since 1900 in spite of the fact that roads, e.g., the Rock Island operating in other States, have begun to employ them and are paying higher wages."[57]

In 1908, El Paso labor contracting agencies placed approximately two thousand Mexican track workers on American railroads every month.[58] That year the federal government estimated that 60,000 to 100,000 Mexicans entered the country every year.[59] By 1910 Mexicans probably constituted the vast majority of track workers on all the southwestern lines, estimated by historians and social scientists at 70 to 90 percent depending on the season. For example, journalist Carey McWilliams argued that since 1880 Mexicans constituted about 70 percent of the section crews and 90 percent of the extra gangs on the six principal western lines: the Atchison, Topeka & Santa Fe, the Southern Pacific, the Union Pacific, the Denver & Rio Grande, the Colorado & Southern, and the Western Pacific.

Most of these scholars appear to base this number on the 1929 study by Charles H. Hufford. Unfortunately, Hufford did not provide thorough citations, and provided only partial information regarding the Santa Fe and Southern Pacific lines. Hufford's data for these two railroad lines, however, seem consistent with my own findings.

According to Hufford, the Santa Fe employed Mexicans on their section gangs at an average rate of 67 percent and 90 percent for extra gangs. The Southern Pacific employed Mexicans workers on section gangs at the average rate of 84 percent and 90 percent on the "special" gangs. Without providing specific data on the other four railways, he stated that on all six railways Mexicans constituted on average of 67 percent of their sections gangs and 90 percent on their extra gangs.[61]

In 1926, California Governor C. C. Young's Mexican Fact-Finding Committee reported that during peak periods, the six major southwestern railroads employed fifty thousand Mexicans on extra gangs. The Young Committee further reported that southwestern railroads employed between 33,000 to 48,000 Mexican track workers per season. And on these roads, Mexicans comprised 75 percent of both extra and section gang personnel. Beyond the Southwest, sizeable numbers of traqueros found work in Kansas, Oklahoma, Missouri, Iowa, and Illinois.[62]

It should be noted that the seasonal nature of track work and the inevitable economic downturns caused these numbers to fluctuate dramatically from one season to the next. These fluctuations are most dramatically seen with extra gangs, i.e., those gangs that railroads hired seasonally and transported to wherever the railroad needed them most. Hufford noted that the Santa Fe normally employed 10,579 men and this figure climbed to 13,667 during peak demands. For the six roads total, the railroads employed from 31,560 to 39,905 on regular section gangs, and 13,104 to 22,923 men for extra gangs.[63]

In 1928, one railroad official for the Chicago, Milwaukee & St. Paul Railway in Illinois reported that "probably not over 500 of our 3566 track employees are Mexicans." Another official for the Chicago, Rock Island and Pacific Railway reported that "on many of our divisions Mexicans predominate in our extra gangs because of the difficulty in securing men locally for temporary service."[64]

The Bureau of Railway Economics reported that data received by the Interstate Commerce Commission in 1911 showed the downward fluctuation among the country's maintenance-of-way employees due to the economy. Between April 30, 1910, and April 30, 1911, the number of track workers decreased 23.3 percent. It further noted that trackmen suffered the greatest reduction followed by shopmen. Despite increased mileage across the nation, "the number of employees per every 100 miles fell during the year from 777 to 707, or 9.0 per cent [*sic*]."[65]

A glimpse at California, one of the fastest growing states in the West and the destination of thousands of Mexican immigrants, indicated that in 1928 Mexican laborers numbered 10,706 on the payrolls of six large interstate and interurban railroads in California alone. Further, as one study noted, "Mexicans have secured a strong foothold in the industries of the State and are certainly displacing other immigrant races and the native-born."[66] More precisely, both interstate and interurban railroads preferred to hire less expensive Mexicans—instead of white men—to work the tracks.[67]

During World War I, the retention of Mexican track workers became a greater concern for both the labor contracting agencies and the railroads due to the increased number of rail shipments and the resultant heavy wear and tear on the tracks. In order to help retain their Mexican labor, some companies ended their relationships with these employment agencies and indeed tried to provide what they thought were better and more suitable living conditions for Mexican laborers compared to those provided by the agencies. To railroad companies like the Santa Fe the most appealing aspect of Mexican labor was their so-called willingness to work hard for low wages. For those workers who provided loyal service to the company for six months to a year the railroads typically offered free transportation to and from Mexico. Traqueros were also allowed to have their families accompany them in some cases, and were provided free boxcar housing for as long as they were employed by the company.

With this context in place, the following chapter examines the day-to-day work life of the traqueros.

chapter 3

work experiences

IN THE 1920S, MEXICAN ANTHROPOLOGIST MANUEL GAMIO collected various corridos from Mexican immigrant workers with the idea of capturing their hopes, fears, preferences and disappointments. As a form of oral history and Mexican working-class culture, corridos reveal much about life among the Mexican workers in the United States and in particular the daily work experiences of traqueros. Gamio recorded one corrido called "Los Enganchados" ("The Captured Ones"), which went this way:

Con los picos en las manos
Nos pusimus a tampar.
Unos descargaban rieles
Otros descargaban "tallas"
Y otros de los compañeros
Echaban de mil malallas.

With our picks in our hands
We set out to tamp
Some unloaded rails
And others unloaded ties,
and others of my companions
threw out curses.[1]

The title of this corrido refers to a form of labor contracting that often characterized the first experiences of Mexican workers in the United States. This corrido recounts track work and some of the various duties that trackmen performed: unloading rails, ties and tamping the roadbed. And, of course, it reflects the physical exhaustion and hard-core vernacular that often accompanied heavy masculine work. Indeed, the most salient issue emerging from this corrido is the collective experience of Mexican trackmen in cooperation with one another in one of the largest employment sectors in the country. And as such, these workers derived overwhelmingly from the Mexican working-class. According to the late Ernesto Galarza, "as a group [the Mexican immigrant worker] they represent the most authentic transplant of Mexican working-class culture in the United States."[2]

This chapter examines the work experiences of Mexican track workers (primarily men) based upon both the attitudes of railroad managers about Mexicans and Mexican attitudes about their back-breaking work on el traque. It describes what they did off the job, i.e., during lunch, evenings, weekends, holidays, as well as how they coped during down-times, unemployment. Specifically this chapter discusses the concept of common labor, working conditions, the effects of Americanization and Taylorism. Finally, it reconstructs the recreational and casual activities of Mexican track workers. In so doing, this chapter shows how traqueros shaped the world of track work within the context and limitations of industrial capitalism through their bonds and relationships with one another as well as with the institution of the railroad.

Mexican railroad workers during the late nineteenth and early twentieth century are perhaps best understood as constituting that part of the American working-class known as "common" labor (or casual labor).[3] According to David Montgomery, common laborers worked in gangs and tended to be male, immigrant, single, and the lowest paid. Common laborers were set apart from the skilled, white, and native-born workers who enjoyed superior wages as yardmasters, roadmasters, superintendents, station agents, inspectors, machinists, mechanics, brakemen, carmen, firemen and section foremen. Railroads also reserved better paying jobs for foreign-born white workers over Mexicans and African Americans.[4] Moreover, common laborers "exchanged simple physical force for a daily wage, whose level changed only gradually over the course of the nineteenth century."[5] Inherently seasonal, temporary, and extremely mobile, common labor has also been the most difficult group to organize into unions. Often referred to as "hobos" and "tramps," common labor possessed an almost inherent sense of itself as the most invisible and powerless sector of the occupational hierarchy, and shared a common disdain for bosses and the wealthy.

Mexican casual labor provided the railroads with the reserve army of labor that it needed to replenish its work force when workers walked off the job, sought better employment elsewhere, or went on strike. And the abundance of Mexican common laborers allowed railroads to hire or lay off thousands of workers at a time during the periods of economic boom or decline. While the majority of Mexicans performed track work, a few found preferred employment as helpers in the shops and the yards.[6] Table 3.1. shows that the overwhelming number of Mexican railroad workers in El Paso were listed as common laborers in 1920. It also shows their movement into skilled railroad occupations. El Paso's proximity to the border explains this occupational mobility. Indeed, some probably learned their trade on one of the Mexican railroads before crossing the border.

TABLE 3.1
OCCUPATIONS OF SPANISH-SURNAMED RAILROAD WORKERS IN EL PASO AS LISTED BY THE 1920 CITY DIRECTORY

Occupation	Persons with Spanish Surname	Occupation	Persons with Spanish Surname
Laborer	188	Handyman	26
Machinist	97	Boltmaker	21
Boltmaker	60	Clerk	20
Machinist	56	Printer's Helper	16
Car repairman	55	Fireman	13
Car repairman helper	49	Gateman	12
Coach cleaner	48	Car carpenter	11
Truckman	37	Machinist apprentice	11
Blacksmith helper	36	Car repairman apprentice	9
Wiper	28	Hostler helper	8
Stripper foreman	27	Helper	8

Occupation	Persons with Spanish Surname	Occupation	Persons with Spanish Surname
Coppersmith helper	8	Special agent	2
Painter	7	Wheel pressman	2
Hostler	7	Airman helper	2
Blacksmith	5	Ash pitman	2
Trackman	5	Craneman	1
Pipeman helper	5	Toolman	1
Stripper	4	Car repairman helper	1
Tinner	4	Hammer boy	1
Supplyman	4	Assistant manager	1
Sweeper	4	Switchman	1
Coppersmith	4	Labor foreman	1
Apprentice	4	Track repairman	1
Tankman	4	Engineer equipper	1
Messenger, telegraph dept.	3	Wiper foreman	1
Subforeman	3	General cashier	1
Painter apprentice	3	Towerman	1
Truckman	3	Machinist Inspector	1
Mechanic	3	Hammer operator	1
Tank helper	3	Brakeman	1
Engineer watchman	2	Drill pressman	1
Fire helper	2	Carpenter helper	1
Stevedore	2	Pipeman	1
Porter	2	Mill helper	1
Oil man	2	Wheel press helper	1
Engineer	2	Airman	1
Engineer inspector	2	Air brake repairman	1
Timekeeper	2	Coppersmith helper	1

Sources: El Paso City Directory, 1920. Mario T. García, *Desert Immigrants* (1981): 67.

Common labor performed track work, that is construction, repair, or work related to maintenance-of-way. Track hands did the most undesirable, dreary and physically demanding work on the railroad, and enjoyed little if any of the romantic qualities of other higher level occupations. Among other things, they dug ditches, shovelled rock, graded road bed, lifted and laid ties and rails, and drove spikes. Traqueros replaced ties, rails, fastenings, frogs (switches) and ballast. Track workers cut the overgrowth of grass and weeds on banks, handled new and old materials, kept ditches open, adjusted fastenings, and maintained gauge (distance between rails) by re-spiking, adzing (trimming ties), raising joints, and ballasting. Then, in addition to all this, traqueros removed snow and ice and repaired the damage constantly being done by water and weather.[7] Depending on geographical conditions and track laying method, it generally required seventy-five to eighty men to lay a mile of track per day. In the winter of 1882, one large Mexican construction crew in Arizona laid a record-breaking forty miles of track in twelve days, averaging three and a half miles per day.[8]

Experienced trackmen possessed knowledge that only the foremen had. Domingo López, for example, veteran track worker from Topeka, understood the effect of temperature on the expansion and contraction of the rails. He also knew that ice would block drainage, and that frost heaved the track. Further, López understood the effect of age on ties, ballast and other metal and masonry materials.[9] Occasionally, veteran traqueros demonstrated knowledge superior to that of their supervisors and foremen.

López recalled the instance when he and his section gang were waiting for an engineer to arrive so that he could insert a frog (a unit of track which allows the train to cross). After waiting for about an hour, he volunteered to "eyeball" the operation since he had installed hundreds of frogs during his long career. His foreman, a white man, questioned the wisdom of installing it without the specifications provided by the engineer, but said to "go ahead" and agreed to back up López when questioned. By the time the engineer arrived, López had the job done. He recalls that the engineer looked puzzled and asked who had installed it. López nonchalantly admitted that it was he. The engineer then measured the track to double check that the frog had been installed properly, to which he said he had never seen anything like it. "It's perfect," he told López.[10]

Skilled traqueros like López typically went unacknowledged by management. Their reluctance to promote Mexican section hands to preferred positions was due in part to the firmly entrenched system of racial and ethnic stratification in the railroad industry as well as in other industries. Important exceptions to this form of discrimination occurred in northern New Mexico and southern Colorado where bilingual Hispanos from New Mexico found employment as foremen of Mexican immigrant section and extra gangs. Until the 1920s, traqueros obtained foremen positions less frequently elsewhere.[11]

Fidel Ybarra's experience in Kansas also betrays the notion that equated track work with unskilled common laborers who took little or no pride in their work. When asked if he had driven every spike in every mile of track in the county, he wryly replied, "Way more than that...I could take you out and show you just about every place I drove a spike."[12] Ybarra estimated that every mile of track had 3,200 ties, approximately 300,000 in the county. He proudly claimed that he worked on every one of them, "including replacing many of them."[13] Despite the years of experience of some trackmen, accidents occurred all too frequently. For instance, in 1903, two traqueros and their white foreman employed by the Los Angeles Railway Company were buried alive on Pecan Street in Boyle Heights. According to a local newspaper, "The laborers were excavating dirt to be used in laying new tracks on First street, when without warning a portion of the high bank, which towered twenty-five feet above them came crashing down."[14] Authorities could identity the Mexican workers only by their employee numbers stamped on a metal tag they wore around their necks. The foreman, Nicholas O'Brien, suffered slight injuries, while the unknown traqueros both suffered broken bones and internal injuries.[15]

The case of Pedro Negrete in San Bernardino, California, likewise demonstrates the danger that skilled trackmen faced. In 1909, Negrete was inspecting the track on a handcar when a locomotive struck and killed him. A local journalist reported:

> It is evident that after striking the man he was crushed beneath the pilot of the leading engine, and then thrown to the side of the track while the velocipede was carried for several hundred feet before being tossed aside. One wheel was found a few feet from where the dead Mexican lay, and his coat...was picked up still further south, torn to shreds. For a number of yards the rails were covered with blood.[16]

While horrible accidents did occur, traqueros more commonly lost a finger or two, smashed a foot or hand, or busted a limb. Ironically, some foremen preferred to hire men who were missing fingers; missing fingers signified experience and presumably meant that such an employee would be more careful.[17]

Track workers injured themselves on the job so frequently that space does not allow for the discussion of the wide variety and conditions under which these accidents occurred. In part, the risks of accidents were advanced especially with the implementation of efficiency programs prompted by scientific managers. Some foremen pushed traqueros to the point where they seriously injured themselves. One such example occurred when trackmen straddled the rail in order to maximize leverage when tightening bolts with a three-foot wrench. The wrench often slipped and so did the traquero, planting the end of his spine on the rail.[18]

In isolated locations, mexicanas found work on the tracks along with men, but railroads paid them half the wages for the same hours and work.[19] Mexican women witnessed their share of accidents on the job as well. An excruciating account of one such accident is provided by Nellie Quinn. During the Mexican Revolution, many young mexicanas became accustomed to traveling and fighting alongside their husbands and lovers as soldaderas (revolutionary women). And, like thousands of other Mexican families, they worked alongside their men on the tracks.[20] Such was the case with Nellie and Francisco Quinn, who shared a boxcar with their two children and their relatives Glafiro, Inez, their two small children, and Doña Sabina (Francisco's mother).

Nellie recalled feeling at first like it was not really work, she said, "I hadn't thought of it as a job. I thought we were just out on a picnic, but the work wasn't hard and I knew we could use the money, so I agreed." She also kept her own wages and managed their boxcar home. Nellie speaks of her sense of unity with her young husband: "We were really getting to be husband and wife by then." She continued, "I loved working hard beside Francisco. When we came home at night we were both tired and hot. We understood each other's pains. That is almost more important than sharing each other's happiness."[21]

Nellie, Francisco, and Glafiro became a track gang unto themselves. Nellie worked for wages by holding the spikes with pinzas (pincers) as the men drove the spikes with martillos (spike mawl). She recalled:

"One day, a very hot day, I almost fainted in the sun. During the lunch break I wasn't able to eat a bite. Up till then I'd always been able to keep up with the men, but that day my womanhood showed up."[22] Instead Nellie fell asleep and the men commenced driving spikes without her. She recalled:

> With one hand Francisco tried to hold the pincers that held the spike and with the other hand he swung the hammer. It was almost impossible to do it. The hammer was very heavy. You needed two hands to lift it. Even then, after a few minutes it weighed a ton. He worked like that for an hour or so, then I guess his arm got tired. He reached down to straighten out the spike with his hand. Glafiro couldn't stop the swing of the hammer and drove the spike into Francisco's hand.[23]

The excruciating scream woke Nellie. She saw Glafiro hunched over Francisco. They loaded Francisco on the hand cart and returned to their boxcar. Meanwhile, Francisco quietly moaned as he cradled his mangled hand. That night Nellie luckily flagged down a passing engine that took Francisco away. Worried and unaware of her husband's whereabouts, Nellie continued working along with other women and men on the tracks; three days passed until someone threw a piece of coal with a note tied around it from a speeding train. It said, "Your husband is doing fine. He is in a hospital in Los Angeles. It is on Washington Street and Central Avenue."[24]

According to the Interstate Commerce Commission, trackmen were injured and killed more frequently than any other sector of the railroad hierarchy. Given the sacrifices that common laborers made, the term "common labor" conjures up a painfully inaccurate picture of the kind of work that Mexican track workers performed.[25]

Track work required skill. Foremen, roadmasters, and other higher ups all recognized the need for experienced track men. Still, they refused to credit track labor as a skillful occupation. In 1915, a twenty-year-old Anglo college athlete argued this point on behalf of trackmen. Having worked as a trackhand, the athlete reported: "I couldn't handle a scythe properly at cutting grass along the road bed. The regular hands laughed at me because I could do only half as much work as they could and the result of my labor was remarkably rough." He continued:

> I couldn't drive spikes. I tried conscientiously enough, but instead of hitting the spike every time I usually hit the rail, battering it badly. I could not "swing" the spike maul properly. The only way I could drive spikes was by taking short strokes from the waist down. Time and again I have seen regular laborers drive spikes with the "pointed end" of the spike maul, which has a diameter of only five-eighths of an inch or thereabouts. And these laborers take a full, free swing with all the recklessness and abandon of a Ty Cobb at the bat.[26]

Certainly the ability to drill holes, tighten bolts, lay and straighten track, among hundreds of other things, calls for a re-definition of common labor.

While the railroad industry as a whole represented complex relationships of interlocking directorates, government leaders and legislation, the industrial hierarchy itself rested on the backs of trackmen.[27] Track workers and their foremen constituted the greatest proportion of the railroad employees. And, while their numbers fluctuated seasonally, they amounted to approximately 30 percent of the work force.[28]

Mexican common laborers must be understood as part of the great body of men in North America of different racial and ethnic backgrounds who, during the nineteenth century, built the canals, railroads, roads, ditches, dykes, mines, and dams, along with municipal construction that ultimately transformed the United States into an industrialized economy. The work experiences of this generation of the Mexican working-class lay the foundation for a greater permanent presence of Mexicans in the United States.

The experiences of traqueros were shaped in part by the attitudes, views, and behavior of white foremen and bosses toward Mexicans. These attitudes are virtually identical to the pejorative images of other working-class and poor immigrants who arrived during the late nineteenth and early twentieth century, except that racism loomed larger for Mexicans and other non-white track workers.[29] Railroad officials, like industrialists, generally valued Mexicans only for their apparent willingness to accept the most oppressive employment for the lowest wages.[30] Railroad documents, trade publications, newspapers, and townspeople in general came to understand that the main reason railroads hired Mexicans over white workers was their willingness to tolerate conditions that few other workers could or would withstand.

Racist attitudes toward Mexicans emanated from the East and South due to the prior experiences of white America with American Indians and African Americans. Of course, anti-Chinese sentiments also stemmed from this legacy of racial oppression. Traqueros' apparent "docility," and their willingness to work under oppressive conditions for low wages without complaint was by far the most widespread racist attitude held in common by track foremen and railroad authorities.[31] The comments of a Santa Fe official in Fresno, California, exemplified this attitude:

> I cannot think of any other class of labor that will stand in the blistering sun for ten solid hours, day after day, without complaint and swing a tamping pick for the small wage they receive. They will work at wrecks in blistering sun and blizzard and washouts in water and mud up to their necks and never shrink from dangerous places, if the foreman will lead the way.[32]

Another observer stated that the Mexican "makes no demand for conveniences, facilities, or arrangements for his well-being, accepting a box car with wooden bunks and no windows."[33] In Chicago, one official reported that: "No we've heard no objection to the Mexicans, altho [*sic*] some of them are as dark as the negroes."[34] A Santa Fe roadmaster had the following to say about the suitability of Mexican labor on the tracks: "Mexicans are better adapted to track work than the other foreign laborers we are getting at present."[35]

The *Railway Age Gazette* featured an article on the usefulness of Mexican track labor. It reported:

> The Mexican is very loyal to the foreman who commands his respect. This was shown by the way the Mexican gangs worked during the unusually severe snow blockade on the Santa Fe and other western lines last winter. Although unaccustomed to cold and snow, they labored more faithfully than many native gangs in opening the lines, frequently remaining on duty more than twenty-four consecutive hours in extremely severe weather.[36]

While this quotation is a gross generalization and overstatement, it reveals the complexity of perceptions among white foremen about Mexican labor.

The following description of Mexicans at work provides a window into the thinking of the general white population in New Mexico. According to an observer:

> 200 Mexican laborers [were] bunched together laying track just south of the flour mill. It was a blistering hot day and the water boys, of which there were about half a dozen, were leading the strenuous life, attending to the thirsts of the swarthy, grimy, sombreroed men who were laying the rails.[37]

An Arizona newspaper summed up another common theme in racist attitudes toward Mexicans in the Southwest, it stated: "As a matter of distinction the life of a Mexican is of no more moment than that of a mongrel dog."[38] Still another example is provided by a track foreman who speculated that the term "greaser" originated from the way insults "slide off the Mexicans like water off a duck's back."[39] Presumably, the ability of Mexican workers to tolerate verbal abuse from foremen also made them uniquely suited to track work.

In 1925, a study of Mexican immigration into Texas also concluded that Mexicans were uniquely suited to industrial labor in the Southwest. It reported:

> They know the art of subsisting in the desert, whether it be obtaining water from hidden sources, kindling a fire under adverse conditions, curing meat and preserving food, cooking with improvised implements, constructing a hut with mud and sticks and thatch, or making use of medicinal herbs to alleviate their ills—in all they are wonderfully adapted to the peculiar conditions under which they must live as laborers in the development of the Southwest.[40]

Consistent with the demand of railroads for "cheap labor," this study reasoned that Mexican workers and the conditions of labor in the desert Southwest were somehow naturally suited to one another. Interestingly, this assumption was not dredged up to argue against their employment in colder regions of the country. On the contrary, the widely held stereotype was that Mexicans worked without complaint under any conditions.

Railroad managers also arrived at the racist conclusion that Mexican workers had childlike minds and required extra patience and instructions in order to perform satisfactorily. In 1911, for example, a Santa Fe Railroad official admonished white foremen to "Never allow them [Mexicans] to argue with you. A great many [white foremen] make the mistake of thinking them stupid. In some ways they are." He continued, "A peon's mind is very much like that of a nine or ten year old child, with the difference that the child's mind matures and the Mexican's never does."[41]

A slightly less venomous railroad engineer had this to say:

> The peon is very much like a child. He likes brightly-colored things and will spend his last cent to get some trivial ornament. He has no thought of the morrow and little does he care as long as his wants for the present are satisfied.... So in handling this class of labor, the foreman must not forgot that by nature the Mexican is a child, and that patience and kindness with firmness and constant supervision are the requisites by which to secure satisfactory results.[42]

Another railroad official suggested that "Mexicans be handled like a bunch of overgrown boys."[43] James P. Craig, a Santa Fe materials foremen in San Diego added that

> Constant reiteration of rules and orders is absolutely necessary in handling Mexicans. Like children, they soon forget. When properly handled they are willing to do a great deal for a man, often working for ridiculously low wages or giving the very best of service for common wages. [44]

Some railroadmen viewed Indian track workers in similar terms. A writer for the *Santa Fe Magazine* said: "Sometimes their conduct appeared the result of childish mischief, and again it seemed the viciousness of the untutored savage."[45] As revealed here, railroad officials tended to lump Mexican workers together as a group and attribute gross generalizations to them collectively. However, as will be shown, these statements also contradicted themselves. But for the moment, the above quotation can be seen as a rather backhanded acknowledgement of how much the railroads exploited Mexican labor. Given the twisted logic of some railroad supervisors, this fact helped to convince whites of the racial inferiority of Mexicans who allowed themselves to be exploited in this manner.

Some supervisors who had the most direct contact with Mexican laborers also promulgated the racist stereotype that Mexicans were naturally dirty and accustomed to rudimentary living conditions. It rarely occurred to these foremen that people at the bottom of a capitalist economy did not have immediate access to bathing facilities, or that the work conditions that the railroad placed them in prevented such amenities. For example, one official stated matter of factly that

> Their manner of living is very crude. Tortitas [*sic*], chille [*sic*] and beans, potatoes, eggs and coffee constitute their principal articles of diet. Camp facilities for them do not require very much, consisting principally of bunk cars, stoves for heating purposes in the winter, and water.[46]

In a trade magazine article entitled "Using the Peon on Track," one resident engineer wrote: "The Mexican peons are not a cleanly people, but still there is not as much sickness among them as one would think. They have no fear of smallpox and seem to want their children to get this disease while young." He continued: "For food they use a great deal of coffee, flour, beans and potatoes and if they can get these they seem to be satisfied."[47]

A labor supply agent concurred with this pejorative view of Mexicans, writing, "As a rule, the average Mexican laborer and family are satisfied with the most primitive arrangements so far as comfort goes; comfort seems to be a remote consideration with him as compared with kindness and fair treatment." He continued, "He is dirty according to the standards of other classes of laborers, but only in a different way, and in this respect has the advantage of desiring to be clean, and will keep so if given very little encouragement."[48]

This attitude stemmed both from the simple ignorance and the necessity to perceive Mexicans in such racist manner in order to retain and reinforce social boundaries between Mexicans and white laborers. On closer examination, however, this perception is far more complicated. Some low level managers for the Santa Fe Railroad often admired and praised Mexican work habits and expressed genuine appreciation for Mexican culture and attempted to assist their Mexican employees both personally and at work. Regional distinction also determined the degree and severity of racial oppression.

White foremen in Texas earned the worse reputation for violence against Mexican trackmen.[49]

On the other hand, railroad managers usually withheld such assistance until traqueros demonstrated years of loyalty to the company. This does not mean that traqueros and white foremen did not socialize with one another and become good friends. To the contrary, many white foremen, especially those of Irish Catholic background, often married Mexican women, and became bicultural. To be sure, railroad managers also concerned themselves with the fraternization between their white foremen and Mexican track crews. One M. Ganley, a yardmaster in Topeka, commented: "white men after they learn to speak the Mexican language . . . want to sit around the Mexican camp with the Mexicans and smoke cigarettes."[50] Certainly, among those white foremen and railroad officials who bothered to ponder the character of Mexican workers, it is probably safe to say that their perceptions of traqueros contained contradictory assumptions. To be sure, railroad officials generally did not want to insult Mexican trackmen or tolerate foremen who severely mistreated their Mexican workers. In 1912, the *Santa Fe Magazine* captured the essence of this attitude, in a grammatically incorrect poem entitled, "Como Estamos Usted" ("How Are You"):

> The Greaser is poor and downtrodden,
> He works for un peso per day,
> But nevertheless he is cheerful,
> With his "como estamos usted?"
>
> He works where most others would perish,
> And he certainly earns all his pay,
> He is cussed, he is fired, but he always
> Has ready, a "como estamos usted?"
>
> No matter how hot or how frio [cold],
> He will greet you most every mañana [morning],
> With a "como estamos usted?"
>
> He's disliked by most all the gringos,
> But I think you'll agree when I say,
> That with all his faults he is cheerful,
> With his "como estamos usted?"[51]

The value of Mexican labor to American railroads and the diaspora of Mexicans on the tracks so impressed one locomotive engineer that he commented: "The Mexican is the 'power behind the pick and the shovel' in the Southwest and west and his influence is advancing gradually toward the Mississippi country."[52]

Like any workers under similar circumstances, traqueros did not passively accept the oppressive conditions, and besides attempts at unionization and formal organizations, traqueros interpreted and employed various strategies to cope with track work in their own terms on a daily basis. And while some Mexican workers harbored negative attitudes toward their own countrymen as well as resentment toward abusive white foremen, the common experience of track work on American railroads created subaltern forms of resistance and solidarity and working-class culture that virtually all Mexican immigrants experienced at one time or another for varying durations. As Paul S. Taylor asserted, "Employment by railroads on maintenance of way, work in round-houses, etc., was by far the most common experience of the emigrants."[53]

As suggested in the previous section, the reports of various white foremen and engineers discussing the characteristics of Mexican labor consisted mostly of gross generalizations, but they, along with other sources, also provide unique insights to the day-to-day experiences of traqueros on the job. Upon closer examination we can see that Mexican workers shared a host of complex situations and responded to them in equally complicated ways. Among other things, they expected fair treatment, tried to ignore insults, "soldiered" (pretended to work), slowed their pace of work, feigned ignorance, joked and talked with one another during work hours, ridiculed foremen in Spanish, shared information about work conditions elsewhere, argued with their foremen, punished foremen and enemies by withholding information about looming danger, and walked off the job.[54]

Perhaps the first and most general misconception about Mexican labor by railroad officials had to do with the fact that trackmen, especially extra gangs, frequently consisted of first generation traqueros. Unless they happened to work as trackmen in Mexico, many if not most had never worked on the tracks before. As verdes (greenhorns) they arrived in the United States railroads not knowing the language and customs; therefore, they appeared ignorant to racist foremen. The dress and mannerisms of Mexican immigrants must have also seemed strange to some foremen who had little or no previous contact with Mexicans.

Recently arrived *peónes* often wore the *pantalón blanco* (white cotton shirt and trousers) and were half-starved and weak. One engineer described the recent arrival this way:

> Usually it takes a couple of weeks to get him "fed up" so that he is able to do a day's work. Often he reaches this country with a straw "sombrero," sandals, and a suit of clothes resembling pajamas. It takes weeks to get him fitted with an American outfit—shoes, overalls, shirt, jumper and a hat or cap. His native hat and sash are usually the last to be changed to conform to his new surroundings.[55]

A Santa Fe roadmaster's observation is consistent with the above. He said, "When Mexicans are first received from old Mexico they are weak on account of being underfed, but after they have been here a month or so they fill out and get strong enough to do the work."[56] Another AT&SF roadmaster reported:

> When arriving on the job he is fatigued by long travel after having suffered previous privations and an insufficiency of food. For a week or two he will seem a poor laborer, but if given a chance to gain strength and is taught his work with patience (Mexicans are slow to learn) he will be an efficient and loyal employee.[57]

Paradoxically, Euro American foremen complained on the one hand that Mexicans were "childlike" and slow to learn, but simultaneously praised traqueros' first season of employment on the railroad. For example one foreman noted:

> The Mexican laborer's first year is the best part of his service for the railway. After he becomes Americanized he begins to reduce on his work. After the Mexican gets so he can talk good English and smoke a pipe he is not of much use.[58]

First-year traqueros worked hardest to impress foremen and thereby retain their jobs. Experienced traqueros determined that under certain conditions it was unlikely that they would be laid off, especially when the demand for labor was high. Also, after the first season of heavy back-wrenching track work, Mexican workers sought and found more remunerative employment in other industries. Indeed, some like Mercedes and Anita Aparicio worked their way from Zacatecas to Colton, California, specifically to secure work in the Colton Portland Cement plant. Others, like Francisco and Nellie Quinn, found themselves in Los Angeles unable to continue to do track work.[59]

Experienced traqueros recalled their own ignorance about doing track work. Some traqueros pretended to not know or understand their foreman's instructions and expectations. By feigning ignorance, traqueros saved their energy. On the other hand, recent arrivals might have been truly ignorant about track operations, but also believed that better employment waited for them elsewhere. And, based on their previous occupational experience in Mexico, they tolerated the temporary conditions associated with track work. For example, by the 1920s, Mr. Silva, a baker in Mexico, worked on the track temporarily before finding better work in Chicago. He listed track work just above picking beets as least desirable jobs.[60] In most cases, mexicanos expected fair treatment and employed any variety of strategies to protest unfair treatment.

Generally, the poorest peónes from Mexico might have tolerated abusive conditions longer than others, while traqueros with railroad or mechanical skills (or the few who previously held white collar positions) learned in Mexico soon sought out better paying and less strenuous work.[61] Verbal abuse especially for first generation traqueros probably went ignored for lack of understanding for the first few days, but traqueros did not tolerate verbal abuse for long. L. J. Hughes, Division Engineer for the Rock Island Railroad had this bit of advice for track foremen:

> The Mexican will stand more abuse than any other class of laborer, and many ignorant foremen think that they are handling a gang of Mexicans effectively by asserting authority and shouting orders accompanied by a stream of strong language. However, it will be noticed that such a foreman is getting no work from his gang and the men will "soldier" as much as possible.[62]

The fact that traqueros withheld their labor power, or "soldiered" in passive response to verbal abuse, reveals a subaltern form of resistance and opposition. Presumably, this kind of behavior would also soon be followed by verbal confrontation between traqueros and the foreman, a complete work stoppage, or walking off the job. Indeed, as shall be shown in Chapter 4, Mexican trackmen often dropped their tools in solidarity with one another in the face of an abusive foreman.

M. Ganley, of the AT&SF, had this insight into the abusive behavior of some foremen: "Mexican gangs will not permit a foreman to curse and abuse them, and where a foreman does this, if they remain with him at all they do very little work. As a rule they will not remain with him."[63] James P. Craig, also of the Santa Fe Railroad admonished foremen:

> It is poor policy to make a practice of bullying these laborers, who are in no position to defend themselves. If this is done the company will have to pay more money to hold a gang or will get only inefficient workers.[64]

Craig's assertion that Mexicans were in no position to defend themselves is not entirely correct. Elsewhere he reveals that traqueros severely tested the foremen's patience and some quit. According to Craig: "If a Mexican can see that he is succeeding in worrying or aggravating you he will in many cases do his best to make your life a burden. Many a good man has been worried off the job in just that way."[65]

Mexicans learned to ignore abusive foremen, especially those who continually berated them, and this had the effect of "pissing-off" the foreman even further. Craig advised grouchy foremen to

> Keep your temper and remember that verbal chastisement will only slide off the Mexicans like water off a duck's back... If you are grouchy all the time the Mexicans soon become accustomed to it, and it ceases to have any effect on them other than to make them dislike you.[66]

Clearly, Mexican trackmen occupied the lowest level of the railroad hierarchy and as such possessed the least institutional power, but they did possess a modicum of power on the roadbed (the shop floor for traqueros). If verbally abused, they consciously dragged their feet, argued with the foreman, or simply walked off the job. Once in a while Mexicans successfully made a Euro American foreman so miserable that he quit. Indeed, in extreme cases mexicanos physically attacked or killed the foreman (see the next chapter). Traqueros also targeted *patroncitos* (strawbosses) for reprisals.[67]

In order to increase "efficient" communication between English-speaking white and Spanish-speaking Mexican trackmen and ease "friction," the Santa Fe Railroad formally published a handbook with "Mexican Translations" for common words used by foremen on the tracks. It noted that the "pronunciations are Mexican, differing somewhat in many cases from the pure Castilian."[68] This qualification suggests that traqueros (or *patroncitos*) assisted with the compilation of this glossary. Craig further advised monolingual foremen:

> A knowledge of the Spanish language, or at least enough of it to be able to explain work, carry on a common conversation and tell what the men are talking about, is absolutely essential in your efforts to avoid friction. If you don't understand Spanish they pass their time ridiculing you to your face. This has a tendency to destroy their respect for you.[69]

Mexican trackmen no doubt enjoyed joking and talking about the foremen within earshot; such talk was probably perceived by other workers as especially brave or risky depending on the circumstances.

Another common way Mexican trackmen resisted complete subordination on the roadbed involved imagination and fun. At times traqueros feigned ignorance when foremen gave them orders, especially if the traquero felt tired and the foreman was disrespectful. Craig, having much experience in this regard, warned his readers, "If you are not on the alert he will take advantage of you by pretending he does not understand. If you will let him, he will lead you into an argument simply to take your attention from the work in hand." Craig continued: "In dodging work . . . they are past masters and will tax the ability of anyone who tries to keep them at their tasks."[70] Notwithstanding the times when traqueros genuinely misunderstood orders, some foremen automatically distrusted Mexicans and expected them to "soldier" as often as possible.

Mexican track workers also shared information about particularly bad or good foremen and job locations. As discussed in Chapter 2, some traqueros returned annually to the same line, division, and section boss rather than chance working for a bad foreman. Indeed friendships developed between some foremen and the men, and in many cases traqueros returned with a relative or friend hoping to place him on the same section or extra gang. And as discussed in Chapter 2, railroads willingly provided company stationery to Mexican employees and virtually ordered them to write home in order to encourage other Mexicans to come and work for that line.

Trackmen generally shared information among themselves about job opportunities and past experiences with various railroads and bosses. The information they shared became what one foremen referred to as a "clearing house" for Mexican track workers. Very hard foremen who drank frequently and abused his men were known through most railroad camps and the lines that employed such men found it difficult to find traqueros willing to work for them. Conversely traqueros withheld useful information from their enemies, and used this information to offer favors and make friends.[71]

Among themselves, traqueros controlled the camp culture and negotiated for space and self-respect. Veteran track workers chose to share their knowledge of track work with some verdes and not others and in so doing exerted influence over who worked unnecessarily hard, which bosses dressed down, and who retained jobs.[72] Again, Craig provided some insight on traquero roadbed culture. He stated:

> If they like a new man they will help him and tell him about his work. But if they are not well impressed with the newcomer they let him strictly alone. They will not warn him if they see him in danger, nor will they aid him if he performs his work incorrectly or works under a strain when there is no necessity for it.[73]

Trackmen often took pride in their work, and guarded the knowledge that they possessed. The work itself required good health and tremendous full-body strength; therefore, traqueros generally decried physical frailty and weakness.

Indeed, the exertion required for six months of track work turned away many strong young men after one or two full days of work. One notable example was Martin Ramírez, who was born in Jalisco, Mexico, in 1885 and who died a famous artist in 1960. Track work almost killed him. He became disillusioned with life in America and developed schizophrenia. Institutionalized in a madhouse, he became a famous artist.[74] The details of Ramírez's life will probably remain a mystery, but his years of work as a section hand likely contributed to his demise. In any case, other mexicanos prided themselves for being able to withstand the oppressiveness of heavy manual labor associated with track work.

Mexican workers expected to be treated fairly. Those foremen who physically or verbally abused traqueros found it impossible to extract a full days work from their gang. And the most flagrant abusers found their men walking off the job without explanation, or asking for their time.[75] In situations where families accompanied the traquero on the job, their first priority was family cohesion. And when a foremen unwittingly discharged one of them, the whole gang would quit.[76] James P. Craig noted that:

> They invariably travel in pairs, trios or groups, consisting of relatives, neighbors or compadres. The different members of these groups will stick through thick and thin, right or wrong. The boss generally will find that it pays to keep track of these groups, as any trouble with one is likely to be followed by demonstrations from his friends.[77]

Oddly enough, foremen expressed surprise when individuals and groups of traqueros suddenly walked off the job in the direction of the nearest town. This action was perhaps one of the most common expressions of traquero protest. Moreover, mexicanos understood that their departure would leave foremen in the lurch and thereby indirectly punished him for his abuses.

Unlike traquero families who tended to live in company-owned section houses or boxcar communities year round, single men or solos most often lived in work camps, bunkcars, boxcars, or cattlecars, fitted with stoves and makeshift bedding. Traqueros considered jobs that required year-round service in the yards to be choice jobs because they could find housing in the nearby barrio and return home on a daily basis. Unfortunately for most solos, the railroad reserved these choice positions for Mexican trackmen with families, and in so doing encouraged and reinforced traditional household relations. In contrast, extra gangs spent weeks or months far away from family and friends.

In the late nineteenth century, boxcar housing was not as widespread as in the decades after the turn of the century. And during the summer season traqueros—both solos and those with families—slept outdoors or built makeshift camps near the railroad yards. Nonetheless, the major difference between solos and married traqueros with families was that the latter ultimately formed the nucleus of Mexican barrios, especially in those areas not previously settled by Mexicans. As shall be shown, this process of barrioization (community formation) cannot be understood outside the context of changing employment patterns and policies of the railroads.

In the 1880s and 1890s, many Mexican track workers built *chantes* (shanties) along the tracks throughout the Southwest and Midwest. Traqueros in South Colton, for example, constructed these makeshift houses with scrap materials discarded by the Southern Pacific Railroad. Scrap sheet metal, tin strips, wood panels, posts, and different size boards were in abundant supply around the railroad yards.[78] Rather than pay rent, track workers merely built their own shanties with these materials and prepared their meals over an open fire. And while uncomfortable, this rudimentary form of housing served as a temporary shelter against the extremes of heat and cold in lower desert regions of Southern California.[79]

A Santa Fe old-timer provided the following glimpse of early grading work and the way in which housing was arranged. He wrote:

> The housing problem was settled by whatever means happened to be at hand. Some had tents; some built dugouts or shacks from old lumber, and many lived in covered wagons of the prairie schooner type, while some resided in bunk-houses, the beds of which usually were infested by hordes of hungry greybacks and bed bugs.[80]

Sleeping quarters for track workers improved somewhat when the railroads used bunkcars, although the bug-infested bedding and extreme climatic conditions continued.

The practices of the commissary gave rise to serious resentment against the exorbitant prices they charged for foodstuffs, work clothes and gloves, among other things. For example, a railroad foreman who hired Elias Garza for $1.50 per day ordered him to purchase supplies at the commissary or lose his job; he bitterly recalled that the railroads "exploited us without mercy in the Commissary [*sic*] camp, for they sold us everything very high."[81] Some commissaries provided workers with meals, while others provided workers with food for their own preparation. And still other comisarias provided both. In some large railroad yards the company provided commissaries for all of their employees, including single workers who lived in boxcars in the yards.

Track work disadvantaged solos in particular; this work compelled them to stay out on the jobs for weeks or months at a time. Once on the job, they had little choice but to complete the work stint. And of course those familiar with this work often refused to accept it. One labor contractor described the difficulty of retaining Mexican track workers. He said:

> I found they would quit after 15–20 days, because they had no cook stoves for tortillas. I supplied the stoves. But still there were not enough [of them willing to stay].[82]

The experience and disposition of Luis García—a 24-year-old solo—toward track work and housing is revealing. García worked for two years on the South Shore Railroad at River Rouge, Michigan, but he quickly grew tired of railroad boarding and bad meals. He said:

> I want track-work, but not where I have to be boarded. I want to board myself. I want tortillas and frijoles and chili. I don't like white beans, and biscuits not done enough; they make me sick. I don't like roast beef.[83]

Another traquero complained:

> We came here on November 23, 1926; 17 solos in box cars. But [t]he box cars had bed bugs and were cold, so I got an apartment.[84]

Clearly single men lived and worked under the most uncomfortable conditions compared to those of married traqueros with families living on semi-permanent railroad housing. And, certainly by the 1920s, railroad officials modified their views regarding the willingness of Mexican track workers to accept the most oppressive conditions. For example, some railroads initiated programs to improve the living conditions for temporary track workers. The Elgin, Joliet & Eastern Railroad publicized its improved labor camp as a means of increasing efficiency and company loyalty among the men, it reported that:

> The movement toward improved housing facilities has been brought about through a realization on the part of the railroads that the money spent would be returned manifold in the better condition of the men in body and spirit, reflected in turn in increased output of work and reduced labor turnover.[85]

Otherwise, as mentioned previously, first-time traqueros learned after their first season that track work provided the least appealing housing and meals. And, therefore, it appears that the majority of seasonal traqueros did not return to the same railroad and division the following year, especially if better pay and more favorable living conditions could be obtained.[86]

Track worker camps varied in size and the degree of comfort, but not too widely. In fact track worker camps were not too different from other types of common labor work camps. Other industries—mining, smelting, agriculture, livestock, and lumber—provided some form of temporary housing for their workers. Laborers' quarters were typically referred to as "bunkhouses," "tie-houses," "labor camps," and or simply "workers' quarters." More importantly, there appears to have been a certain physical similarity to work camps. Rexford Tugwell's description of a lumber camp where boxcars were used to house the men, could just as easily been that of a track workers' camp. According to Tugwell:

> The bunk-houses are usually unlighted and unventilated. Bunks are of wood and likely to be lousy. Men sleep on boards—in some few camps now there are steel bunks with wire springs. There is the inevitable air-tight stove in the center between the rows of tiered bunks around which at night wet woolen clothes are hung. The heat is intense and the sweaty, steamy odors of a bunk-house at night would asphyxiate the uninitiated.[87]

In 1914, railroad camps averaged twelve to sixteen men per car, and most railroads separated men by race, that is, white men and non-white. Railroads allowed 170 cubic feet of air space per man, which according to one study was half of what government statute required.[88] In most cases, the only ventilation came from the large doors in the middle of the car and when closed, bad air could not escape. According to another observer:

> The stench on the inside from grease and filth is almost unbearable. The [boxcar] dwellers work, eat and sleep in the same wearing apparel, and apparently never change a garment until it is worn out and another secured.[89]

This investigator described a track laborers' camp in Wisconsin this way: "Cess pools for waste water were rarely found and where holes had been dug they were open and overflowing...the conditions described above apply mainly to white men's camps. Foreigners camps as a rule are worse."[90] Camps like these generally attracted swarms of flies and vermin, and railroads attempted to control illness by dismissing sick men and sending them back to the city.

Obviously under such conditions, many if not all of the workers experienced tremendous stress. Indeed, not unlike opium use among Chinese track workers, traqueros occasionally consumed cheap whiskey, "loco weed," and other substances in order to dull the senses. Without heterosexual release, Mexican workers, like all workers under similar conditions, may have engaged in homosexual activity, although if so, it was probably not widely accepted and kept discrete.[91]

In 1920, social workers like Tugwell were interested in bringing attention to this dimension of life among America's working-class men who lived under these conditions. He described the anxiety of life this way:

> And there is nothing else but a sodden, unrestful sleep in the evil-smelling air; except perhaps the talk in the bunk-house may turn for a little while to women—it seldom gets far from sex—talk that injures and bruises the natural urgings in men, perverts them, cuts them loose like coiled snakes in the infrequent intervals of vacation in town.[92]

Aside from the semi-religious association between "evil" and "uncoiled snakes" referred to in this quote, it is a telling reflection of the fear of sexually deprived men on local townswomen. Indeed, it seems that social workers, bosses, and public officials feared that the living conditions among large numbers of single, womanless men would arouse homosexuality and sexual perversion.[93]

Through the employment agencies, railroads routinely deployed men to various work sites; according to Carlton H. Parker: "without blankets or even sufficient clothing... they are forced to sleep packed together for the sake of warmth."[94] Parker further concurred with an investigation that purportedly showed that "there are social dangers which a group of demoralized, womenless [*sic*] men may engender under such conditions of greater menace than the stereotyped ill effects of insanitation and malnutrition."[95]

Parker arrived at the following conclusion regarding the insalubrious impact of the common laborer population:

> There is here, beyond a doubt, a great laboring population experiencing a high suppression of normal instincts and traditions. There can be no greater perversion of a desirable existence than this insecure, undernourished, wandering life, with its sordid sex expression and reckless and rare pleasures. Such a life leads to one of two consequences: either a sinking of the class to a low and hopeless level, where they become through irresponsible conduct and economic inefficiency a charge upon society; or revolt and guerrilla warfare.[96]

Of course the advent of Americanization institutionalized this perception of "casual laborers" with post-World War I programs. Radical labor organizations like the IWW provided further impetus to Americanizers to convert immigrants into 100 percent Americans for fear of class war.

Track workers frequently contributed to the dangerous conditions within their communities. In 1903, Arcadio González died of a stab wound in the chest as a result of an argument stemming from a card game in a traquero encampment near the area of Verdugo Canyon outside of Glendale, California. The unnamed assailant also sliced at another worker by the name of Félix, but Félix grabbed a shotgun from his tent and chased the assailant away. A few days later, traqueros fought among themselves at a christening at the Southern Pacific boxcar colony near the Los Angeles River. Again, the incident involved alcohol, knives and guns. A local paper quoted a track worker: "As one of the cholos [lower-class Mexicans] said, there was no motive—it was just a fight."[97] While we may never know the real motives behind the violence, trackmen in general were not known for their genteel behavior. Yet, this kind of violence was not strictly a racial or cultural feature of Mexican immigrants.

In January 1909, Jesús Valdez and his fellow workers customarily celebrated payday with a few drinks at a local cantina and continued to imbibe at their tie-house owned by the Phoenix and Eastern Railroad in Florence, Arizona. Valdez and his compadres drank into the wee hours of the morning and were apparently passed out, when one of the men knocked over a stove or lamp that ignited the tie-house they slept in. The fire quickly spread and all the men except for Valdez escaped from the burning shanty. Neighbors discovered the blaze at about four o'clock in the morning, and found only a few fragments of his body.[98]

chapter 4

Labor struggles

On April 24, 1903, a dramatic scene took place on Main Street in Los Angeles when more than thirty Mexican women (primarily the wives of strikers) confronted several dozen *esquiroles* (scabs) imported from El Paso by the Pacific Electric Railway Company (PE). Owned by Henry Huntington, the PE attempted to replace striking "cholo" laborers represented by a new Mexican union, La Union Federal Mexicana (UFM). Huntington arranged for police to arrest any picketing Mexicans. The mexicanas harangued the esquiroles to join the strike and marched boldly onto the grade site and wrestled shovels, picks and tamping irons away from the hands of the strikebreakers.[1] *The Los Angeles Times* referred to the mexicanas as "Amazons" from various parts of "Sonoratown," the principal Mexican settlement.[2] Onlookers, mostly Mexicans and Anglos, stepped over and around railroad ties, rails, wheelbarrows and mounds of dirt as they strained to watch the commotion. Within moments some esquiroles joined the strikers and others fled while a few others argued with the women and the striking traqueros and futilely tried to defend their actions.

While the UFM sought a basic wage increase and overtime pay for weekends, their strike symbolized a rare moment when Mexican women organized for their class interests and simultaneously pushed the culturally defined boundaries of proper women's behavior to challenge the power of Huntington on behalf of their families and community. It is equally significant that the 1903 strike is one of the first attempts at labor unionization among Mexican track workers in the United States. The reasons traqueros were slow to unionize are varied and complex; however, the specific context of Mexican employment on both railroads and streetcar lines is important. Indeed, the 1903 PE streetcar strike was not the first expression of militant traquero resistance to the ubiquitous power of the railroad.

As was shown in the previous chapter, mexicanos resisted complete domination in unique and creative ways. Subaltern forms of resistance occurred far more regularly than attempts at unionization. Nonetheless, traquero unionization represents the formal expression of class consciousness as well as racial and ethnic pride. And as such, the organization of Mexican track workers for their community and class interests represented a higher form of political struggle, albeit one that possessed its own internal contradictions such as the gender oppression of women in the household. The demands of trackmen tended to reflect the traditional organization of the family, for example, the control over the conditions of men's work, wages, hours and access to family housing. For traqueros, traditional family arrangements remained intact. Moreover, traqueros actually preferred an arrangement where railroads allowed families to accompany them in boxcar housing for seasonal track work.

This chapter examines various formally organized attempts by traqueros to exert control over the conditions of their work and livelihood. The first example is the militant direct action tactics of Las Gorras Blancas of New Mexico during the 1880s and 1890s. The next section examines the first organization of traqueros in the United States in the region of southern Colorado and northern New Mexico. It will then analyze the innumerable strikes in rural outlying regions where Mexican trackmen worked and laid down their tools for any variety of reasons, except for unionization. Such spontaneous strikes contrasted dramatically with the attempts at protracted strikes organized in urban areas where traqueros who worked on interurban street railway systems typically struck for better wages, shorter hours, and at times union recognition.

Labor organization among Mexican track workers, whether mutual aid associations, fraternal lodges, or formal unions, must be understood against the general status of organized labor in the United States and Mexico. In general, both American and Mexican labor unions in the late nineteenth century still resembled secret societies and fraternal organizations. In Mexico, Euro Americans dominated the leadership and the rank and file of the most powerful railroad unions.[3] At best, railroads in both countries tolerated trade associations among skilled native-born white men, and simultaneously rejected the notion that they must share power with workers.[4]

The Knights of Labor and the IWW represented the only national labor unions to address the needs of Mexican railroad workers north of the border. In 1916, the IWW launched the Railway Workers Industrial Union, which sought to organize all railroad workers (skilled and unskilled) into one big union, as opposed to unions organized along trade lines represented by the railroad hierarchy. However, as will be shown, government repression during World War I ended the Wobblies' effectiveness. However, the Brotherhood of Maintenance of Way Employees, founded in 1887, successfully organized Mexican and Black trackworkers in significant numbers, but not until the late 1920s.[5]

The most successful labor organizations among railroad workers tended to be those of engineers and other skilled white and native-born workmen.[6] Locomotive engineers possessed far more influence and collective power over the railroad industry than did any other sector of railroad employees. Conversely, Mexican trackmen, like most trackmen, collectively represented the most powerless sector of the railroad employees. Notwithstanding engineers' strategic importance to the industry, railroads violently opposed unionization among engineers and absolutely loathed the idea of unionization among its lesser skilled employees. Thus, the objective position of traqueros in the overall structure of the railroad employment hierarchy limited the possibility of effective unionization among Mexican workers.

The earliest moment of significant labor organization among traqueros is thus limited to the few instances where they organized on interurban railway systems. The urban locale allowed traqueros to build community support and carry on strike activities. Aside from this kind of organized resistance, the militancy of Las Gorras Blancas in New Mexico during the late 1880s and early 1890s appears to be the only form of organized class-conscious resistance that preceded attempts at unionization among Mexicans.

Called the "white caps" by the authorities, Hispano horsemen in northern New Mexico adopted the name Las Gorras Blancas. They are well known to students of southwestern arid borderlands history for their night raids on horseback, fence cutting, and violent confrontations with Euro American businessmen, ranchers, land speculators, and the Atchison, Topeka & Santa Fe Railroad. However, historians have been reluctant to make clear linkages between the working-class motives of Las Gorras Blancas and the Knights of Labor. Alienation from their land, and not alienation from their labor, has been seen as the primary motive behind the activities of Las Gorras Blancas.[7] The violent resentment of Hispanos directed toward the railroad stemmed from the combination of dispossessed land and the alienation they experienced as second-tier trackmen whose labor was exploited by "outsiders" through a dual wage system. The historical intersections of these forces can first be examined by a discussion of Hispano dispossession from the land.

With the possible exception of south Texas, Hispano land tenure differed fundamentally from other Mexican settlements in the Southwest. First, land tenure in New Mexico and parts of Arizona and Texas stemmed from Spanish and Mexican land laws. Hispanos lived on ejidos (individual plots of land) and shared forest land surface and subsurface water rights communally. No individual possessed rights to that communal land above that of the village and therefore could not sell off portions. In the 1870s, the expansion of the railroad into the Southwest brought new pressures on traditional land rights among Hispanos.[8] The causes and development of the legal battles surrounding land grants like Millhiser v. Padilla, have been examined thoroughly elsewhere and need not be repeated here.[9]

Nearly three centuries of Spanish and Mexican rule produced various legal traditions and forms of land tenure in northern New Mexico. Spanish and Mexican land laws differed in several ways from Anglo American laws, not the least of which involved the form and quality of record keeping. Hispanos in the northern frontier, like most frontier peoples, did not possess a tradition of highly efficient record keeping. Thus, Hispanos found themselves severely disadvantaged when they attempted to prove land ownership in American courts. Underlying this discontinuity of privilege lay a chauvinistic supposition that most Hispano land owners possessed little or no right to land in the first place. As with other parts of the annexed portion of Mexican territory, gringos regarded Hispanos as outsiders. Thus, through the courts, subterfuge, intimidation, and violence, Anglos wrestled away choice agricultural and grazing land. This dispossession in turn gave rise to a body of Hispano men and women who then found themselves compelled to find a new way to survive economically.[10]

The demand for labor from local villages accompanied the advance of the Santa Fe Railroad through southern Colorado and northern

New Mexico. Hispanos for the most part reluctantly complied with the new demands of the American era. On the one hand, they needed to work to survive, and on the other hand, the Santa Fe Railroad provided them with opportunities for employment as trackmen, teamsters, and other unskilled kinds of employment.

As has been shown earlier, Hispanos in southern Colorado were among the first Mexican origin workers to find employment as track hands. Logically, Hispanos first entered into social relations with the railroad hierarchy as laborers and attempted to accommodate the changes brought about by the railroad. And while Hispanos performed wage work prior to the arrival of the railroad, the magnitude of wage labor employment opportunities must have appeared endless. Unfortunately for most Hispanos, railroads and mining operations reserved the best paying and most secure occupations for Anglo men.

As the railroad advanced through and around traditional communal lands, Hispanos responded in various ways. A few entrepreneurial Hispanos sought to profit from the new economic arrangements, while others sought redress in the courts to protect their rights. And of course, many others merely sat idly by as the conflict and changes unfolded. However, a significant minority of Hispanos (from 700 to 1,500 men) clandestinely led by three brothers Juan Jose, Nicanor, and Pablo Herrera, armed themselves and dressed in long coats, slickers and white caps to conceal their identity. The White Caps were not unlike other nineteenth-century secret societies that sought to redress injustices brought by land speculators, "monopolists and capitalists," as well as local politicos who collaborated with them.[11] And, consistent with rural peoples unaccustomed and unfamiliar with formal judicial procedures and laws, Las Gorras Blancas relied on armed struggle and direct action.

The strength of Las Gorras Blancas in the same New Mexico towns where Terrence V. Powderly's Knights of Labor also held large memberships suggests that the two groups held mutually reinforcing loyalties. Indeed, oldtimers in New Mexico recall that "Las Gorras Blancas" was just a nickname since they preferred to be called Los Caballeros de Labor (Knights of Labor).[12] Like the Knights of Labor, Las Gorras Blancas raised concerns about the unequal wages paid to Mexican track hands. For instance some traqueros reported that Las Gorras Blancas encouraged them to demand equal wages with white trackmen. Hispano teamsters working for the AT&SF Railroad reported that Las Gorras Blancas attempted to force them to strike for higher wages at gunpoint.[13] In another case, Las Gorras Blancas left a written warning on the wall of the AT&SF section house at Bernal. It stated, "To all section foremen and operators. You are advised to leave here at once otherwise you will not be able to do so. Given under our hands to the White Caps."[14] Thus it would seem that Las Gorras Blancas clearly understood the relationship between the dispossession of their land and the concurrent process of proletarianization. Indeed, they articulated this consciousness through their public pronouncements. On the one hand, they claimed that the lands belonged to the people and at the same time demanded equal pay for their work. [15]

The activities that New Mexico lawmen attributed to Las Gorras Blancas included kidnapping, murder, robbery, and destruction of property. In many cases, authorities never actually produced evidence to convict the White Caps. Indeed, as with social banditry in other parts of the Southwest, Euro American authorities attributed virtually all unsolved crimes to Las Gorras Blancas.[16] However, among their many activities, Las Gorras Blancas cut and destroyed barbed wire fencing, burned barns and haystacks, split railroad ties and telegraph poles, set flame to AT&SF Railroad trestles, and ripped up railroad tracks with axes.[17] In one notable case of mass arrests and indictments, a New Mexico judge released forty Hispanos when the prosecuting attorney could not produce his star witness. One angry Las Vegas official described what happened afterwards:

> Immediately upon their discharge these accused, with several hundred other Mexicans assembled in the Plaza near the court House, and there for three hours were harangued by members of their crowd, some shooting pistols in the air, and all in the most excited way, denouncing the public officers and the laws.[18]

This public assembly and demonstration of mass support probably caused state officials to fear the possibility of an even greater "Mexican" uprising.

Among the various expressions of protest in which Las Gorras Blancas engaged, fence cutting appeared to be a preferred form of protest. Fenced land represented a basic contradiction to the centuries-old tradition of open grazing and communal access to the land.

After Las Gorras Blancas destroyed a fence a second time, they warned large white landowners and certain wealthy Hispano farmers to not rebuild the fence or "prepare for eternity."[19] In July 1890 one such translated notification is included in the papers of Territorial Governor L. Bradford Prince. It reads:

> This notice is with the object of requesting you to coil up your wire as soon as possible from the North and South sides. They are fences which are damaging the unhappy people and we request you further to coil up your wires as soon as you can to the agricultural land, and if you do not do it, you will suffer the consequences.
>
> Sus Serbidores [Your Servants]
> los gorras Vlancas [*sic*].[20]

Warnings like this one convinced many white landholders to comply. Even the Las Vegas sheriff, Lorenzo Lopez, complied with the order.[21] In December 1889, Las Gorras Blancas issued a manifesto entitled Nuestra Platforma which explained their grievances and declared their plans in both Spanish and English. In this document, Las Gorras Blancas declared themselves the defenders of the rights and interests of the people, especially the "helpless class," which in New Mexico clearly referred to the masses of the Mexican people. Las Gorras Blancas declared their moral authority over corrupt politicians, monopolists, lawyers, the judicial system and political leaders. They further declared themselves "law abiding citizens." They warned the authorities, "be fair and just and we are with you, do otherwise and take the consequences."[22]

Consistent with the official government response to radical labor movements and populist struggles elsewhere, the Las Vegas District Attorney, Miguel Salazar, privately described the platform of Las Gorras Blancas as "anarchical, revolutionary, and communistic."[23] According to Salazar, three captured members of the White Caps confessed that someone from Utah or Colorado had organized "such societies under the name first, of White Caps, but now under the name of Knights of Labor."[24]

Las Gorras Blancas in particular targeted the AT&SF Railroad probably for its symbolic association with the general transformation of their culture, and specifically for exploiting their labor as trackmen.

For example, in one incident they destroyed 9,000 railroad ties on Santa Fe property.[25] And in another incident they "cut in two pieces" six thousand railroad ties contracted by a local Hispano politico for the Santa Fe Railroad. Ultimately, the AT&SF Railroad cancelled contracts for railroad ties from the county which resulted in a loss of approximately $100,000 per year to San Miguel contractors.[26]

As a secret society, Las Gorras Blancas did not keep a roster of members, thus making it virtually impossible to cross-check names of White Caps with AT&SF payroll records. It stands to reason, however, that many individual members of Las Gorras Blancas had worked as track hands previously. As has been discussed earlier, itinerant track work for the railroad was one of the few wage labor occupations that Hispano farmers could easily obtain. Indeed, names like Herrera, Gallegos, and Martinez fill the payroll ledgers of the AT&SF for the 1890s.[27] Some white track workers and would-be small farmers joined the White Caps; moreover, it is likely that some of these gringo Gorras Blancas also represented some of the rank and file members of the Knights of Labor. Howard Lamar suggests that the White Caps sometimes claimed to belong to Eugene Debs' American Railway Union.[28] In any case, the real or imagined links between these various organizations require further investigation.

In the 1890s, White Cap activity declined, but Hispano trackmen continued to organize. According to Sarah Deutsch, Hispanos founded seventeen *mutualistas* between 1885 and 1912.[29] In 1900, Hispanos founded eight branches of a Colorado mutualist/labor organization exclusively for Mexican trackmen on the Santa Fe Railroad called La Sociedad Proteccion Mutua de Trabajadores Unidos (The Society for the Mutual Protection of the United Workers, or SPMDTU) in Antonito in Colorado. Little is known about the early years of SPMDTU, and records for the budding union have not survived. One of the founders, Celedonio Mondragón, previously belonged to the Sociedad Proteccion Mutua Por Ley y Orden (Society for the Mutual Protection of Law and Order) and thus possessed organizing skills and a keen sense of racial politics in southern Colorado.[30] Indeed, in large part, the new conservative direction of Hispano resistance probably came about as a result of the failed politics of the Las Gorras Blancas and the Knights of Labor.

Like mutual aid organizations in other parts of Mexico and among ethnic groups elsewhere in the United States, the SPMDTU emphasized honesty, morality, family, and obedience to both federal and state laws. According to historian Fredrick C. Sanchez, SPMDTU leaders limited membership to men of "Spanish" descent between the ages 18 and 60.[31] Regrettably, as with other original Mexican inhabitants of the annexed Southwest, Hispanos saw themselves as socially superior to Indians and Mexican immigrants and took pains to remind Anglos of these differences.

Additionally, the SPMDTU, like mutual aid societies elsewhere, developed an insurance program for members as well as charity and patriotic programs. In these culturally conservative associations, women performed only auxiliary and supportive functions. It also borrowed some of the political views of Las Gorras Blancas regarding bossism, political corruption, and racial discrimination. However, unlike the Las Gorras Blancas, the SPMDTU did not engage in armed struggle and guerrilla warfare. [32]

Only a few early mutual aid societies among Mexican laborers transformed themselves into bonafide unions. Instead, they became fraternal lodges. The SPMDTU eventually became recognized by the state of New Mexico and survived as an ethnic fraternal association well into the 1970s.[33] Rather than try to redirect and transform the advancement of Anglo industrialization and political dominance through violent direct action as did Las Gorras Blancas, the SPMDTU acquiesced so long as it could retain a modicum of control over immediate village and community concerns.

Mexican trackmen partook in and sometimes led their own struggles to gain better wages, better conditions, shorter hours, and eventually union recognition. Except for the occasions when Mexican track workers walked out in solidarity with upper echelon railroad workers, they rarely engaged in protracted strikes to the degree that railroad workers in other regions did. Between 1873 and 1874, for example, railroads in the Northeast and Midwest cut wages, laid off workers, and withheld wages.[34] Trackmen, brakemen, switchmen, and firemen of Euro-American background had no unions, but they organized collectively nonetheless. According to Herbert Gutman, most of these strikes occurred in small towns where workers could muster popular support and carry on protracted strike activity. After failed attempts to recapture lost wages and jobs, trackmen on various railroads tore up sections of track, disabled locomotives, spiked switches, and cut telegraph wires.[35]

During the 1870s, the AT&SF had only begun track construction into the Southwest, but the towns the railroad created eventually led to the conditions that paralleled the Northeast. In 1886, significant numbers of Mexican and Black trackmen participated in the Great Southwest strike. However, white workers marginalized their influence.[36] Indeed, maintenance-of-way crews struck for a variety of reasons, and in some cases for slightly different reasons than traditional strikes in small towns. For instance they sometimes struck in order to retain a favorite foreman or to have a mean-spirited foreman replaced. Working on the track, F. Huerta recalled that:

> A rail on our track was found to be one inch from the end of another by the roadmaster, so he fired the American foreman. The Mexicans liked the foreman and struck unless he should be returned.[37]

In this case, the entire gang of Mexicans lost their jobs for refusing to work under a different foreman.

On the afternoon of July 16, 1903, three hundred shopmen and yardmen in La Junta, Colorado, struck against the Santa Fe Railroad because their checks had not arrived the day before. In 1916, Mexican trackmen struck the Wabash Railroad in Peru, Indiana, for wages, and then again for better housing and food. A Wobbly newspaper attempting to shame white workers toward militant action reported that "the Mexicans are more exacting in their demands than Americans."[38] In March of 1908, about 250 traqueros in Colton, California, struck to replace their foreman. In response, the Santa Fe Railroad decided to deport 200 of the striking Mexicans to El Paso. Railroad officials played down the rebelliousness of the Mexican laborers and dismissed their concerns. Later that month, another similar confrontation erupted in the adjacent town with the same results.[39]

Traqueros struck to protest against foreman Johnson at his isolated section house near Cajon Pass north of San Bernardino. However, in this case violence followed. After several weeks of tension, a confrontation ensued between a dozen or so strikers and a Santa Fe Railroad foreman. In the melee, two traqueros, Francisco Bernardo and E. Geraldo, fought with Johnson and stabbed his son in the upper leg. Johnson grabbed a shotgun and frightened off the striking traqueros. Local authorities captured and arrested Bernardo and Geraldo the following morning.[40]

Although few in number compared to trackmen, Mexican shopmen also struck for more wages and better hours. In the spring of 1917, Mexican shopmen working for the Simpson Contracting Company in the Santa Fe shops in San Bernardino demanded an eight-hour day instead of the required ten. They also demanded a twenty-five cent raise over their two dollar per day wages. A few days later, Santa Fe section hands also demanded a wage increase and threatened to strike. It appears that these strikers, both the section hands and the shop employees, belonged to the Asociacion Mutua de Trabajadores Unidos (Mutual Association of United Workers) in San Bernardino. As elsewhere, the mutualista served, for all intents and purposes, as a labor organization and provided a vehicle for Mexican workers to conduct strike activity. The railroads refused to accede to the demands and dismissed the strikers.[41]

Railroad and federal authorities blamed "outside agitators" for the "trouble."[42] Newspapers throughout southern California blamed Mexican revolutionaries and agents of the Industrial Workers of the World in nearby Colton for inciting rebelliousness.[43] These strikes, and an explosion of other strikes in surrounding industries, emerged simultaneously with America's entry into World War I and the fear that Mexico might side with Germany in the war. The upsurge of strike activity at the spring of 1917 had more to do with the seasonal demand for labor in the railroad and citrus industry than it did with radical union activity. However, the convergence of the rising cost of living associated with the war effort and the activities of radical organizations among Mexican workers gave rise to alarmism in southern California and the Southwest.[44]

Traqueros also struck to protest the high cost of food, supplies, clothes, and housing. In addition, at times they struck to protest the dismissal of a friend or relative, or to protest the late or non-payment of wages. Because these kinds of strikes occurred spontaneously and frequently, few have ever been recorded.[45] Unlike strikes that threatened a shift of power, traqueros typically won these lesser strikes.

Professor Emilio Zamora informs us of the frequency of strikes in South Texas where Mexicans and Blacks made up virtually every railroad construction crew. In the early 1900s as the railroads neared the border, independent operators cut wages from one dollar to seventy-five cents per day with little or no concern for the Mexican workers and their families who camped along the line. Traqueros struck only to be replaced with new workers from Mexico.[46]

Despite a long tradition of militancy and strike activity in the nineteenth century, traqueros in the United States failed to effectively organize into unions until the 1920s. An important exception to this trend occurred in Texas where both the railroads and the white controlled unions colluded to maintain second-class status for Mexican railroad workers. As in Colorado and New Mexico, these Texas traqueros formed their own mutual aid society, La Sociedad Ignacio Zaragoza.[47] In 1903 one of the founders of this mutualista, Silvino Rodriguez, returned to Chihuahua, Mexico, and founded the La Union de Mecanicos Mexicanos (Mexican Machinists Union). Zamora has noted how many Mexican labor leaders cut their teeth in the labor movement on the Texas side of the border. Filled with resentment for being treated like second-class citizens in Texas, they returned to Mexico and refused to be displaced by Anglo workers in their own country.[48]

The failure to organize permanent labor unions among Mexican track workers was therefore due to a variety of reasons. Because they were considered itinerant "unskilled" laborers, railroads could easily replace track workers in most cases. And since traqueros performed most of their work (construction and maintenance-of-way) in isolated regions, prolonged strikes were impossible to carry out in the absence of community support. Mexicans also needed to work. Both single and married traqueros came to the United States to work, and compared to wages in Mexico (twenty-five to thirty cents a day) a dollar per day in the United States at first appeared to be an improvement.[49] In part, recent arrivals may have given railroad employers the impression that they were content. Thus, when Mexicans struck, their action appeared to shock authorities who blamed "outside agitators" and typically dismissed Mexican workers' grievances.

Due to the seasonal nature of their work, Mexican trackmen tended not to be interested in striking for better wages or working conditions, and least of all unionization. Unlike the few skilled Mexican mechanics in Texas, most traqueros fully understood their vulnerability to being replaced with scab labor from El Paso or elsewhere. Traqueros themselves had no personal stake in a strike when they lived so far away from home and therefore possessed no form of family and community support. And for their part, Anglo labor unions—even trackmen's unions—had no interest in organizing Mexican immigrant workers due to their own racism against these workers.

Moreover, Mexican track workers most often traveled and worked with other men on an itinerant laborers' schedule. For six-month stints traqueros lived with other men out on the line and visited nearby towns for recreation, usually on Saturdays. These seasonal workers depended completely on the railroad for housing and meals, and hence without family and community support prolonged strikes and unionization efforts were impossible. Finally, labor organization and unionization required men with experience in organizing, including possessing negotiating skills and the ability to feed and support a large body of strikers with reserve funds and or food stuffs. Mexican trackmen generally did not have either the experience or these resources.

The 1903 Pacific Electric strike in Los Angeles stands as one important exception to the apparent lack of class and ethnic militancy among Mexican track workers. In the late nineteenth and early twentieth century, Los Angeles stood as the fastest growing city in the West. The total population of the metropolitan area stood at only 11,183 in 1880. Between 1890 and 1930, however, the population of Los Angeles increased from 50,000 to 1.2 million, becoming the nation's fifth largest city.[50] According to Professor Pedro Castillo, the Mexican population grew as well from 3,000 to 5,000 in 1900, to 30,000 to 50,000 in 1930.[51] In 1900, Mexicans constituted about 15 percent of the total population of Los Angeles. The Southern Pacific Railroad and the Atchison, Topeka & Santa Fe Railroads served as the conduit for this population growth; both railroads contributed heavily to the growth of the Mexican population.

As in other major cities, Mexican immigrants found employment mostly in unskilled and semi-skilled jobs. The railroads employed thousands of men and transported thousands more to various employment sectors, especially agriculture, construction, and mining. On the rare occasions when Mexican workers struck, industry typically ordered replacements from El Paso or other reserve labor pools. Mexican laborers working for Huntington's Pacific Electric Railway earned about one dollar and fifty cents for a ten-hour day. The Huntington lines also required its track workers to purchase their goods and food at the company store where workers paid double for goods available elsewhere. Moreover, because of the mexicanos' difficulty with English, the PE assigned each traquero an employee number stamped on a metal tag which they wore around their neck.[52]

In March and April of 1903, approximately eight hundred Mexican trackmen working on the Los Angeles Railway and the Pacific Electric Company organized themselves into a union called La Unión Federal Mexicana. The local Council of Labor led by Lemuel W. Biddle, affiliated with the American Federation of Labor (AFL), assisted and supported the Mexican union. On Friday morning, April 24, 1903, the UFM presented their wage demands to the PE superintendent: from fifteen cents to twenty cents an hour for day work; thirty cents for night and overtime work; and forty cents an hour for Sundays. The immediate response from railway authorities was uneven. Epes Randolph, the PE General Manager, quickly acceded to the traqueros' demands without authorization from Huntington. The Los Angeles Railway, on the other hand, immediately refused the demand. A spokesman told the UFM, "'I will not give you a nickel, and you can go straight to h—.'"[53]

Receiving conflicting responses, the workers struck. *The Los Angeles Times* described the early moments of the strike:

> The calling of the men from the Main-street job yesterday was dramatic. An agitator ran along the street like a Mercury, ordering the peons to stop work, whereupon hundreds of them immediately dropped their shovels and picks and followed after him like a lot of sheep... they were ordered to meet in a hall down in the depths of Sonoratown, where they were harangued by the agitators during the afternoon. [54]

The traqueros timed the strike to coincide with Huntington's attempt to have the street railway system completed prior to May 6 and 7, 1903, in time for the annual Los Angeles Fiesta celebration. President Theodore Roosevelt agreed to review the parade on Main Street downtown. The UFM believed that it could best pressure Huntington to accede to a wage increase under community pressure.[55]

In an attempt to smash the strike, Huntington ordered all striking Mexicans fired. He then recruited white, Mexican, Black, Italian, and Japanese replacement workers from Long Beach and other outlying areas. The PE also paid strikebreakers two and one-half-cents more per hour than what the UFM had demanded.[56] By the following afternoon, the PE had replaced almost half the strikers. General Manager Randolph spoke on behalf of the PE, and offered this explanation for Huntington's actions:

> Mr. Huntington proposes to run his own affairs and can in no manner accept union dictation. While he is disposed to pay good wages, these peons would never have struck had it not been for Biddle [AFL official] and the agitators.[57]

This quotation reveals the contempt that PE officials harbored toward unions in general and what they loathed to give up: power. As with strikes throughout the country, the issues were not so much the hours to be worked and wages to be paid, but rather who would determine them. Moreover, the attitude expressed was consistent with the chauvinistic perception of Mexicans as "beasts" incapable of organizing without the assistance of white men. Mr. Randolph stated that: "They [Mexicans] are very much like a band of coyotes; they will follow a leader. I venture to say on my own authority that not one-fourth of these men knew what they were quitting work for."[58]

City officials also dispatched hundreds of police to the downtown area of Main Street in order to forestall trouble. No doubt the show of force by the police department for the purpose of preventing strikers from talking to scab workers intimidated traqueros associated with the strike. In order to avoid personal harm and trouble with the police, it is likely that some workers abandoned the strike altogether and sought work elsewhere.[59]

Indeed, it appears that a small number of traqueros broke with the UFM and returned to work. Some may have feared company reprisals, the threat of violence (from strikers or police), or loss of income, or lacked confidence in the union's ability to win. Moreover, it is likely that these men and their families lived in boxcar housing provided by the company and found it too difficult to stay with the strike and risk eviction.[60]

In response, the fledgling union held a mass meeting and plotted strategy to win support from the numerous AFL unions.[61] As a symbol of their ethnic solidarity, the strikers raised the Mexican flag at their makeshift headquarters in the barrio. Local officials interpreted the display of Mexican nationalism as a sign of subversion, rather than as an expression of Mexican working-class solidarity. Traqueros believed that their exploitation as workers was linked to the prevailing racist attitudes Euro Americans held toward Mexicans.[62] According to an A. Nieto, a UFM organizer, the union printed circulars and distributed them throughout the Mexican labor camps in southern

California and as far away as El Paso. Nieto also requested the cooperation of the Mexican government to help prevent the entrance of Mexican strikebreakers into Los Angeles and to avoid the "indignities and abuses" at the hands of railroad authorities.[63]

On Friday evening April 24, 1903, the strikers tried to form a picket line at the Main Street construction site. However, local law enforcement officials prohibited them from doing so because, according to police, the "Mexicans were drinking too much."[64] It appears that law enforcement officials ordered policemen to surround the strikebreakers in order to prevent them from communicating with strikers. Thus, the UFM devised methods by which to talk to the esquiroles in order to convince them to join the strike. The UFM sent individual organizers into various work sites while pretending to be passers-by. At the moment the section boss left the scene, the organizer jumped into the ditch and quickly picked up a shovel and commenced digging along with the strikebreakers. At that moment, the organizer initiated a conversation with the esquiroles and informed them about the purpose of the UFM and the strike. According to a Los Angeles Times reporter, this method proved unsuccessful: "He [the organizer] had just time enough to whisper into one man's ear and a 'company detective' had five digits entwined in the neck of the 'agitator's' jumper."[65]

The most dramatic moment of the strike occurred in the middle of the afternoon on Saturday, April 25, 1903, when over thirty mexicanas marched on strikebreakers. *The Los Angeles Times* reported:

> a squad of women appeared on the street where those who had taken the strikers' places were at work. The women had come from various parts of Sonoratown and all of them were relatives of or sympathizers with the Mexicans who had gone on strike the day before. They had been carefully told what to do; but who gave them their order is not known. The women approached the workers and began seizing the shovels, picks and tamping irons which they were using. [66]

As this quotation suggests, these women may have been instructed by the union leadership to undertake this novel form of direct action. However, it is likely that the women independently decided to march on the strikebreakers in an expression of solidarity with their husbands, brothers, and sons. Mexicanas, as homemakers, mothers, and wives must have also felt that the strike had as much to do with them as with their men folk. Some striking traquero families had been evicted from their company boxcar homes, and despite the oppressive conditions, those without close family and friends nearby had nowhere else to turn. Nonetheless, it is clear that these mexicanas took it upon themselves to confront the strikebreakers and correctly predicted that their action would take the esquiroles by surprise. Soon after, the police arrived in force and dispersed the crowd.[67]

It appears that this confrontation was the single instance when mexicanas of the UFM attempted this tactic; however, Huntington soon witnessed its full impact. Two days later, the total number of traqueros on strike climbed to fourteen hundred.[68] In part, the dramatic increase may have also been due to the arrival of Santa Teresa Urrea, "La Nina de Cabora" the famous Mexican religious leader, folk healer, friend of the Mexican working-class, and enemy of the Porfirio Diaz regime in Mexico. She arrived on Monday, April 27, 1903, with a procession of thirty-eight Mexican women down Main Street. As the procession passed the construction site over fifty traqueros dropped their tools and walked off the job. She led them to the UFM office where she exhorted the virtues of working-class and ethnic solidarity. According to Nieto, the secretary treasurer of the UFM, Santa Teresa arrived at the union's request: "She is assisting us in the strike. The Mexicans will do whatever she says. We have sent word back to Mexico to our relatives not to come to this country to work, as there is a strike on."[69] Her presence doubtlessly had an impact on the heavily Catholic Mexican working-class community of Los Angeles. If nothing else, her presence and public support of the UFM made it unpopular for mexicanos to work for the Pacific Electric Railway during the strike.

Since the first day of the labor conflict, Huntington had refused to recognize the UFM or to acknowledge that the strike had slowed the construction of the street railway. He had immediately hired replacement workers from outlying areas, especially Long Beach, and transported them back and forth daily. And, he had ordered two hundred Mexicans from El Paso, Texas, to replace the striking traqueros. Local newspapers reported the arrival of thirty-five strikebreakers one week and twenty-five the next.[70] With the depleted work force, Huntington's replacements worked day and night to complete the Main Street line. By May 5 Huntington had successfully completed the downtown section of track. All traces of a Mexican trackworker uprising had been erased prior to the Fiesta parade. While the great intensity of the strike spanned less than two weeks, Nieto and Biddle declared the strike ongoing. Nonetheless, it appeared clear that Huntington had effectively crushed the strike.[71]

For their part, the strikers maintained a good degree of discipline, but the union lacked reliable officers. It appears that the elected treasurer absconded with the union's strike funds. The UFM collected a total of $600 in less than a week from membership dues, labor union donations, and the Socialist Party.[72] Additionally, some strikers sought other employment or left Los Angeles altogether. Moreover, the strike had never really materialized in the formal sense because the police prohibited picket lines and protest demonstrations anywhere near the construction site. Thus, the strike failed on its own measure to win a wage increase and it failed to gain union recognition. In June of 1903, the UFM declared that the strike was still on, and, a year later, the UFM struck the Pacific Electric Railway again. On March 1, 1904, the UFM struck to protest a pay reduction and to demand sanitary housing.[73]

The 1903 Pacific Electric strike stands as a good example of the difficulty that Mexican trackworkers faced at organizing themselves into a union and in striking to better their conditions. Indeed, it was one of the most dramatic expressions of Mexican track worker militancy in the early twentieth century. The context and conditions that led to this particular strike and others like it are unique to the urban context of Los Angeles and the presence of a dynamic Mexican ethnic community in particular urban locations. Mexican trackmen with families and nearby homes tended to organize and strike for traditional union purposes while single traqueros, especially in outlying areas, almost never struck with the intention of unionizing. This is not to say that solos never engaged in strikes, but they did so for fundamentally different purposes and expectations.[74]

The conditions on the PE were probably no worse than those on the major railroads. In fact, the conditions of work on the PE probably represented an improvement over the conditions found on the major western roads. At least Mexican urban communities provided a haven for itinerant workers. The fact that the Mexican track workers' union appeared in the urban metropolis of Los Angeles as opposed to the desert Southwest is significant. Major strikes like the 1903 PE strike never occurred in the desert or in other isolated regions where traqueros found themselves.

The essential difference between the PE strike and other southwestern railroad strikes was that it occurred in a highly urban context. Traqueros living near and among other working-class Mexicans used their families and community ties to sustain themselves during the strike, by depending on family and neighbors for support. Additionally, labor unions and other progressive organizations stepped forward to lend support in the form of money and resources to the strikers. Newspaper coverage by the Times and the *Record* also helped to draw wide community focus on the causes, conduct, and outcome of the strike.

Mexican track workers also made new political allies during the strike. Both the *Record* and the Socialist Party of Los Angeles sympathized and offered financial support to the strikers. In an editorial prior to the strike, the *Record* denounced railroad slavery:

> THE RAILROAD TIES OF THE SOUTHERN PACIFIC AND THE SANTA FE ARE WATERED WITH THE BLOOD OF PEON VICTIMS TO CONTRACT LABOR. IT IS A JEALOUSLY GUARDED SECRET. A SECRET THAT THE *RECORD* HAS EXPOSED, AND TO THE SHAME OF ALL CONCERNED, HAS FLUNG NAKED BEFORE AN ASTONISHED WORLD.[75]

In fact, the *Record* ran stories documenting the oppressive conditions that traqueros worked under in Los Angeles. While this journalism may have been coincidental, it is more likely that the *Record* had full knowledge of the impending strike. In addition, the Socialist Party had published investigative articles in its official organ, the *Socialist*, a year prior to the strike (see chapter 5), and members volunteered money and time to assist the families of the strikers.[76]

Finally, given the rather invisible occupations of Mexican common laborers at the turn of the century, the dramatic events surrounding the strike helped to give the Mexican community a new political identity. Contrary to popular belief, Mexican workers were not merely tractable and docile "beasts" and "dupes." Mexican workers, including Mexican women, demonstrated intelligence, ingenuity, and courage when they marched on the gang of strikebreakers and wrestled their tools away. Certainly, the religious and spiritual support offered by Santa Teresa underscored the cohesiveness and ethnic solidarity of the Mexican working-class community in Los Angeles. Indeed, the mexicano community provided an ideal environment for the arrival of Mexican political exiles associated with Ricardo Flores Magón and the Partido Liberal Mexicano.[77]

The phenomenon of racial conflict and the uneven pace of cultural adaptation profoundly shaped the various expressions of traquero working-class militancy. An examination of racial attitudes between people of Mexican background and Euro-Americans may shed some light on this complex issue. Racial attitudes ranged from mutual respect and cooperation to vicious hatred and resentment. And, of course, these attitudes varied from region to region over time. But perhaps the general perception that characterized Mexican-Anglo relations during the late nineteenth and early twentieth century was rooted in racism and Anglo chauvinism. This racism was at first expressed in biological terms, and then later in cultural terms. Chauvinism accompanied racist thinking and expressed itself most often in attitudes of intellectual and physical superiority. It is important to note that racial attitudes among people of Mexican background (Indians, mestizos, criollos) preceded the racism of white North America and partially combined to form a complex form of racism during the late nineteenth and early twentieth century.

In 1880 American-born Europeans and European immigrants were not, in general, interested in understanding the differences between Hispanos, Mexicans, and Spanish-speaking Indians. Perhaps the only exceptions to this disinterest were the very early Euro-American settlers and immigrants who, due to their few numbers, adapted to Mexican and Hispano customs and language often through intermarriage with Mexicans and Hispanos. To Euro-Americans, Mexicans and Indians looked similar and were lumped together and regarded as biologically inferior. And from the perspective of some Euro-Americans, Mexicans were no different from African Americans. In 1928, a Mr. Wuerst, an employer of Mexican immigrants, reported that, "A lot of these damned ignorant Germans called the Mexicans niggers."[78] Indeed, as has been shown by Arnoldo De Leon, some Anglo Texans regarded Mexicans as more contemptible than Indians due to their mixed blood ancestry. In any case, as the reasoning went, Mexicans lost the war of 1848, and refused to make use of the land adequately.[79]

Euro-American characterizations of Mexicans and Indians were strikingly similar; they were typically depicted as lazy, lying, cheating, filthy, shiftless, and worthless.[80] Ironically, however, both Mexicans and Indians were simultaneously recruited in large numbers to perform track work throughout the period under study. In 1909, a writer in the *Santa Fe Magazine* described the Mojave Indians of California as harbingers of "filth, shiftlessness, laziness and general worthlessness."[81] And in 1920, Helen D. Marston, a white social worker in San Diego, grossly generalized about the character of Mexicans: "Everyone knows that the Mexican does not have the same standards of honesty that the American has." She continued, "to tell the truth is not so important a thing among our [Mexican] neighbors as it is among us."[82] Paul S. Taylor collected the following perception of Mexican track workers by a railroad official who echoed Marston's understanding of Mexican culture. The official stated that Mexicans "are not used to any modern methods of living—sanitation, bathing, eating, etc. After they are well fed for a short time they are good laborers."[83] Another railroad official had this to say about Mexican workers' intelligence: "In receptiveness they compare normally with children in the third to fifth grade in our public schools. They learn to do by watching others as id [sic] occasioned in childhood."[84]

The perception of Mexican workers as children was not uncommon; indeed, American Indian workers were also perceived by Euro-American railroad officials in this way. For example, in November 1914, James Craig, a Santa Fe official, published an article entitled "How to Handle Mexican Labor" in which he stated, "A peon's mind is very much like that of a nine or ten year old child, with the difference that the [white] child's mind matures and the Mexican's never does."[85] However, this perception needs to be reconciled with the following contradictory view of Mexican track workers. He continued,

> A great many men make the mistake of thinking them stupid. In some ways they are. In dodging work, however, they are past-masters and will tax the ability of anyone who tries to keep them at their tasks.[86]

Thus, it appears that James Craig was uncertain as to whether Mexican track workers were truly childlike or whether they at times feigned ignorance in order to slow down the work process and save energy. Indeed, work slow-downs have long been an unofficial means by which working-class men and women exerted a form of informal control over the work-place. Some whites also perceived Indians as having childlike minds. A writer for the *Santa Fe Magazine* had this observation about Indian track workers, "Sometimes their conduct appeared the result of childish mischief, and again it seemed the viciousness of the untutored savage."[87] Certainly this perception harked back to the early colonial views of American Indians as the "children of nature" and, at best, the noble savage. [88]

In the nineteenth century, white authorities in New Mexico believed that "the average Mexican is too much like the Indian," and restricted Mexicans from legally trading with Indians. They preferred Anglos trading with Indians since trade between Mexicans and Indians "would not perform its designated function of transforming Indian culture," which was Americanization.[89]

By the 1920s, these racial attitudes had changed slightly. In part, this shift was due to the change from biological to cultural theories of inequality among social scientists. Rather than biologically inferior, Mexicans and Indians were now frequently regarded as culturally inferior. The reasoning of Kenneth L. Roberts, a writer for the *Saturday Evening Post*, typifies this posture: "Anybody with a long winter evening to waste on meditation might do worse than meditate on the difference between people who haven't bettered their lot materially in 200 years."[90] In the same article, Roberts identified the "cholo or greaser class" as the "riffraff of the Mexican race found along the railroads and in the Mexican cities. [91] This facile distinction between good and bad Mexicans was widespread, but is inconsistent enough to reflect the fluidity of ideas and images of and about Mexicans at any one time. The racial/ethnic images and attitudes of Mexicans changed with fluctuations in the economy, but even then there was tremendous variation at the local level. In the early nineteenth century, the village of Agua Mansa, California, was founded by a colony of New Mexican families to help repress local Indians. Some of these families along with the older California elite were regularly distinguished from the recent immigrant or "cholo" who worked for the railroad and lived along the tracks.[92] In 1905, a small Colton paper acknowledged this difference: "[they] make a distinction between a Mexican and a Cholo. The old-time Mexicans of California have the respect of all their neighbers [sic]. The gang of Cholos employed by the Santa Fe have been engaged in the work of changing and lengthening the switches."[93] Thus the railroad recreated and reinforced racial boundaries, and made a distinction between recently arrived "lower" class Mexicans from the older and "better" class of Mexicans. Moreover, since the railroads were the first to employ Mexican immigrants in many areas, it greatly influenced the racial division of labor in other employment sectors. Inter- and intra-racial and cultural conflict evolved concurrently and were often impossible for contemporary Hispano observers to separate. In 1890, for example, *La Voz del Pueblo*, of New Mexico castigated the publishers of *La Cachiporrita*, a Republican campaign paper, for using "race prejudice" to attack José L. López as candidate for sheriff. According to *La Cachiporrita*, López was "nothing else but a dude" who never associated with Mexicans. It continued, López, "speaks such perfect English he has entirely forgotten the Spanish, so he will be ignored on election day. Down with the dudes."[94] Herein, race and culture were deliberately linked in order to reinforce the racial/ethnic boundaries between Spanish-speaking Mexicans and English-speaking Euro-Americans. *La Cachiporrita* attempted to alienate López in the eyes of the greater Spanish-speaking community so as to reinforce Hispano ethnic solidarity. Mexican immigrants and those Hispanos who successfully retained their Mexican ethnic identity looked upon those who denied their Mexican identity with suspicion. Almost forty years later an unfriendly writer had the following observation of the intra-ethnic hierarchy among Hispanos:

> one finds today, in New Mexico, a Spanish American aristocracy that looks down with some distaste on a Spanish American working class. The working class, in turn, looks down with great disfavor on the type of Mexican peon that is now pouring into the United States in such large numbers.[95]

Thus despite a common cultural background and a common experience of alienation by American institutions, both Hispano and Mexican railroad workers did not readily identify with one another in order to exercise racial/ethnic solidarity. In 1928, Ottilie Falcone, a former traquero in Chicago, shared this observation about his fellow Mexicans. "I am not ashamed of being a Mexican. But I am not proud of the mexicans [*sic*] in Chicago. [T]hey are mostly Indians of the lower uneducated classes of peones."[96] Survival, not social change, was foremost in the minds of most working-class Mexicans and Hispanos.

Due to the social degradation of Mexican immigrant workers in the United States and the competition for jobs, American Indians and Mexicans also shared certain hostile attitudes toward one another.

In 1912, "Indian Bob," a shop worker in Kansas, angrily admonished those who assumed he was Mexican and quickly identified himself as a full-blooded Sioux.[97] Spanish surnamed Indians were not uncommon in Arizona and New Mexico. For example, James Pino, Juan Poncho, and Herbert Paisano, all of the Laguna clan, held machinist positions in the Santa Fe railroad shops of Winslow, Arizona.[98] By 1912, in spite of the availability of American Indian labor, the railroads preferred Mexican immigrant workers. It appears that Mexicans virtually replaced American Indians on track work in the Southwest. Racial attitudes toward Mexicans in the Midwest and Central Plains differed from those in the Southwest since Mexicans in the Midwest were more properly identified as "immigrants" (and identified themselves as such) rather than as native-born Hispanos or Mexican Americans. As shown in earlier chapters, virtually all of the immigrant adult male Mexicans in the Central Plains and Midwest during the early twentieth century were railroad workers at one time or another. Mexican women and children joined their men folk soon after Mexican railroad workers found housing or better employment in other industries. By the 1920s, the obvious increase in the numbers and variety of people of Mexican background in the United States made racial/ethnic, regional, and class lines much clearer than they had been in previous decades. Hence the nature of race thinking did not change significantly. Witness the following quote by a social worker, ostensibly sympathetic to people of Mexican background:

> The Mexican present shades of color ranging to that of the negro—although no negro features—to that of the white. The result is confusion. A Mexican girl enters a street car and sits down among the whites and the conductor tells her to sit among the negroes. She refuses on the ground that she is "no nigger." A Mexican worker on a city job where both negroes and white men are employed refuses to drink out of the negro drinking cup and the foreman beats him up. [99]

As revealed in this quote, some Mexicans held hostile or ambivalent racial attitudes about African Americans and themselves. While it is impossible to determine exactly how widespread this kind of attitude was, it would not be far off the mark to suggest that virtually all people of Mexican background held similar views at one time or another, this attitude being both a product of white racism inherent in North American history and that of Mexican colonial history.[100]

As such, Mexicans and African Americans appeared to have shared a competitive position relative to occupations on the railroad, with Mexican workers possibly enjoying a position only slightly higher than that of black workers on certain railroads. In 1928, one South Chicago foreman offered the following racial/ethnic hierarchy:

> The succession of races here has been Swedes, Irish and Germans; Poles and Slavs; negroes, Mexicans. The young Poles now don't want the hand work; they want to start at the top. The Americans are the same way. The Mexicans, by accepting the hand labor are preparing themselves for the better positions later.[101]

This racial/ethnic arrangement was also influenced by the fluctuations in the economy, and the related availability of Mexican immigrant labor. During the early 1920s the Illinois Steel Company hired blacks and white casual laborers as strike breakers. "Neither were satisfactory," according to Thomas McKenney, an Irish foreman, "So I got the Mexicans from the southwest. The foreman liked them from the first, because of the unsatisfactory experience with the negroes and whites. They say 'Give us some more Mexicans.'"[102] A white employee at Sawyer's Biscuit Company in Chicago had this virulently racist observation about Mexican and African American women:

> Our Mexicans are white. Yes, some of them are dark, just like the niggers. I wouldn't want to live among them. I want to live among white people. No, there's no prejudice against them. Some niggers try to pass themselves off as Mexicans where thet [*sic*] know Mexicans are employed. About 15 were here the other day. They knew a few Spanish words. But no niggers will get by me.[103]

Another instance where Black workers attempted to pass themselves off as Mexican immigrants was recorded by Paul S. Taylor. While standing in front of a labor agency in Chicago, a number of black men attempted to pose as Mexicans in order to get jobs picking beets, only to be informed that these jobs were for real Mexicans only.[104]

The fact that African Americans attempted to pass themselves off as Mexicans suggests that there was probably closer competition for casual labor jobs between Mexicans and African Americans than between either group and Euro-Americans. It is possible that some Mexican immigrants with African features may have been mistaken for Blacks and denied employment on that basis. In either case, the intersection of race, gender, and class revealed here seems particular to the Midwest. However, further examination of this seemingly regional peculiarity is needed before useful generalizations can be made. Racial attitudes held by Mexicans toward Blacks and other groups were mixed and do not form a clear pattern of attitudes about any one racial/ethnic group. Nonetheless, it appears that racial attitudes among Mexicans toward other racial/ethnic groups is best understood as horizontal racism (relations between non-whites toward other non-whites) as opposed to vertical racism (relations between white and non-white). Notwithstanding the hierarchical arrangement of racial groups in the United States, the fact that Mexicans possessed racial (not to be confused with racist) ideas and were aware of racial inequality in the United States is also a reflection of the racial attitudes and racial bias in Mexico and among Mexicans in the Southwest. Whether in Mexico or the United States, the complex interplay of racism, capitalist oppression, and cultural alienation (especially in the United States) reinforced a racial/ethnic and class hierarchy in the railroad industry with only relatively few exceptions.

Due primarily to the racism in American society in general, and the racist policies of American unions in particular, the unionization of Mexican track workers occurred very late in the history of the American labor movement and by that time, technological advances in track maintenance required fewer track workers. In 1877 white track foremen organized a foremen's union, which ultimately became the Brotherhood of Maintenance of Way Employees. Like the other big brotherhoods at the time, they refused to organize non-white and the foreign-born into the union. Although sources are scarce, it appears that the first unionized traqueros appeared in the early 1900s among bilingual Hispano foremen from the region of northern New Mexico and southern Colorado.[105] Between 1905 and 1919, the Industrial Workers of the World also made a number of attempts to organize "gandy dancers" (track workers) in the West, but to no avail.[106] With the exception of the UFM, the IWW represented the only national union to attempt to organize common laborers. However, the postwar Red Scare brought an abrupt end to the IWW and its impact on Mexican trackworkers was minimal. The history of traquero labor struggles was an uneven process shaped by multiple factors. The contours of these struggles depended on the pace of capitalist development, social and geographical context (rural/urban), the concentration of the Mexican origin population, traditions of protest, and political leadership, among other things. The first traqueros in southern Colorado and northern New Mexico formed Las Gorras Blancas and the SPMDTU in direct response to the transformation of their homelands and culture. Later, immigrant traqueros in Los Angeles organized the UFM which appears have been the first attempt at formal traquero unionization in the United states. Both Hispano and immigrant traqueros engaged in countless and unreported low-intensity strikes for a variety of immediate reasons. And, as shown, mexicanas took active, albeit subordinate roles in these strikes.

Notwithstanding these few important expressions of ethnic solidarity and class struggle, the overwhelming majority of traqueros never involved themselves in the violent activities of groups such as Las Gorras Blancas or unionization struggles. Mexican trackworkers more commonly engaged in passive and subaltern forms of resistance. Nonetheless, the attempts at organization by Las Gorras Blancas clearly demonstrate that Hispano trackmen wanted desperately to be treated equally. And, as Mexican immigrants streamed into the United States and quickly outnumbered the native-born Californios, Hispanos, and Tejanos, the immigrants occupied the lowest rung of industrial jobs. In terms of railroad jobs, the majority of Mexicans found themselves relegated to track work. Therefore, Las Gorras Blancas, the SPMDTU, and the UFM can therefore be understood as some of the earliest and boldest attempts at Mexican labor organization and unionization among one of the most powerless sectors of the labor force.

chapter 5

Boxcar communities

CLEOFAS CALLEROS, AN OLD-TIMER AND RETIRED SANTA FE depot official in El Paso, recalled the numerous Mexican track workers and section hands "whose groups of houses dot the desert from here to Los Angeles."[1] Indeed, as this chapter shows, traquero houses dotted not only the southwestern desert, but the entire line from El Paso to Chicago. These dwellings constituted the beginnings of many Mexican immigrant communities in the United States. Community and family formation is clearly tied to Mexican industrial employment on the railroad.

This chapter examines the origins, variety, and social conditions of Mexican boxcar settlements and community development. It argues that while Mexicans could be found throughout most of the railroad occupational hierarchy, most worked in seasonal track work. As such they lived in a variety of company-owned housing. Boxcar communities probably represented the most common form of housing for Mexican workers and their families. It also demonstrates the linkage between this process and the rise of scientific management regarding Mexican section hands. This chapter concludes that railroad paternalism gave rise to the Mexican boxcar settlements during the late nineteenth and early twentieth century as a means of controlling section hands.

Generalizations about the life in boxcar camps are next to impossible to arrive at given the diversity of experiences across the country. Similarities in housing did not give rise to broad similarities in social experience among Mexican track worker families. And while economic hardship and social inequality were widespread, Mexican immigrant men, women, and children interpreted and resisted the conditions they faced by drawing upon their immediate surroundings and their experiences in Mexico. In fact, what is striking about Mexican railroad housing in boxcar communities and other company-owned housing is the diversity of experiences. And while we see clear patterns emanating from boxcar settlements along the various lines, such patterns were not identical throughout the United States.[2]

Railroad worker communities and boxcar settlements are best described as multiple processes of "barrioization."[3] For example, by the 1910s Mexican railroad workers found housing in boarding houses or in the homes of friends and relatives in sprawling urban centers such as Chicago and Los Angeles. Moreover, in regions once settled by Indian, Spanish, and Mexican pueblos especially in New Mexico and Colorado, the pattern of pueblo and village settlements certainly preceded railroad boxcar communities. Notwithstanding the devastating social impact of the railroad and Mexican immigration into these older Hispano settlements, it must be stated that these boxcar communities were typically not inhabited by Hispanos but rather by immigrants. Hence the Mexican immigrant experience is indeed almost exclusively associated with boxcar communities. To be sure, some Hispanos lived in boxcar housing, but typically as itinerant workers and not with wives and children.

More typically, boxcar communities tended to include families, while bunkcars and bunkhouses tended to house individual male track workers. There were occasions when Mexican immigrant families actually accompanied their men while working on the track, but this form of track labor housing was more similar to bunkcars that housed single men. However, the presence of women and families in boxcars or work encampments differed fundamentally from the daily life experiences of solos, as I have discussed in Chapter 3.

The origins of Mexican boxcar communities and Mexican railroad worker settlements in the United States emerge from both the tradition of track work in the United States and *rancherías* in Mexico. North of the border, for example, the Santa Fe railroad's policy shift regarding their retention of Mexican section men gave rise to boxcar communities in the Central Plains and the Southwest. Immediately after the completion of the first transcontinental railroad, Irish and Black workers increasingly found more remunerative employment as "common laborers" in other industries. Over the fifty-year period discussed here, the ebb and flow of this labor shortage combined with the fluctuations in the economy stimulated successive waves of immigration from Mexico, with the greatest increase occurring between 1910 and 1930.[4]

European immigrants first inhabited the earliest boxcar communities in the East and parts of the Midwest, and the same was true for African American track workers in the South.[5] Hispano families in southern Colorado and northern New Mexico probably lived in the first boxcar communities in the Southwest. Towns like Las Animas and Trinidad, Colorado, preceded the arrival of the railroad, and then grew as a direct result of their proximity to the railroad. In the 1860s, an early railroad surveyor noted that Trinidad was "the first and only Mexican town to be found north of the Raton Mountains."[6] In the 1870s Hispano villagers helped lay the first ties and rails through this territory and lived in boxcars. In the late 1870s, Antonito, in southern Conejos County, "was the toughest town in the state" according to an engineer for the Denver & Rio Grande Railroad.[7] He recalled that Antonito "was a railroad town with lots of Mexicans and Indians."[8] In 1881, a newspaper in Raton, New Mexico, published a directory of Mexican railroad workers who lived in boxcars.[9]

In 1880, T. J. Peter, a mining developer, purchased a large portion of land immediately south of Starkville, Colorado, from the original Hispano inhabitants of that region with the promise that he would allow them to retain ownership of their homes. Most of the Hispano men worked either for the Trinidad Coal Company founded by T. J. Peter himself or the Santa Fe Railroad. Conflict emerged between the Hispano workers and the mining company when it attempted to expand its operations and encroach on Hispano lands. In a letter to the Santa Fe's president, Allen Manvel, a company official said:

> They have been living on the lands ever since, and have increased in number. It is the custom among the Mexicans for a son, when he gets married, to build his house close to or adjoining that of his parents; it is quite likely, therefore, that such portions of this sub-division as it is now thought best to deed free to the Mexican occupants make up a parcel of land somewhat greater than was originally promised by Mr. Peter.[10]

The Hispano landholders attempted to win their deeds through the courts, but lost to the mining company. And while the Trinidad

Coal and Coke Company intended to grant the deeds to the Hispano families, the Santa Fe Railroad prevented them from doing so. This process of displacement represents only one of the many ways that American capitalism pushed Hispanos into the ranks of common labor.

South of the border, Arthur J. Corwin has argued that the origins of Mexican migratory workers lay with *rancherías*, as Spaniards referred to them, that emerged on the social margins during the colonial period. Rancherías consisted of small semipermanent roaming communities composed of semitribal groups or extended families which became "squatter camps" on the outskirts of old haciendas, mines, and plantations, or wherever these people could find seasonal employment. Corwin argues that the rancherías and squatter camps of Sonora produced the "first migratory workers in the American borderlands during the California gold rush, during the mining boom in Arizona in the 1860s and after, and during the construction of the Southern Pacific Railroad in the 1880s."[11] Aside from the simplicity of the argument, Corwin provides no substantive data to support the assertion that rancherías became "squatter camps." Moreover, he does not account for the non-ranchería Hispano track workers who found railroad employment in Colorado in the 1870s as has been thoroughly discussed in Chapter 2. Nonetheless, these two trends represent the convergence of cultures from which railroad labor and housing patterns developed.

With railroad construction into the Southwest, boxcar settlements and other track worker camps became almost exclusively associated with Mexican track workers. The AT&SF probably initiated this pattern of housing for Mexican track workers. Other major lines extending into the Southwest, like the Southern Pacific, Chicago Rock Island and Pacific, Union Pacific, Baltimore and Ohio, the Chicago & Alton duplicated the new Santa Fe policy.[12] Before this important policy shift, Mexican track workers usually came as solos and either accepted some form of company housing or found housing on their own.

This railroad policy regarding the employment and housing of track labor varied from one region and from one rail line to the next and represents the uneven process of capitalist development. In 1903 railroad officials in South Texas could not obtain sufficient boxcars to house their Mexican "Peon" laborers, and opted to build tents on the tops of flat cars. According to an observer, "Here the Mexican family groups hived together ... with those old flat cars bedecked in their white canvas not unlike gigantic hooded goblins." Officials for the St. Louis, Brownsville & Mexico Railroad aptly termed this form of Mexican housing as a "rolling village."[13] An eyewitness described the scene this way:

> One of the decorative features ... was the long cords radiating from these camp cars to the mesquite trees, from which there were suspended long strips of raw, red beef and goat meat drying in the sun. The surrounding bushes were usually embellished with freshly washed clothes. The smell of cooking pervaded the camp, and around the family clung the faint and abiding flavor of garlic, blended with the odor of corn shuck cigarettes. They had stones on which to grind their corn and earthen pots for cooking *frijoles* and meat stews, over fires built from twigs and brush out on the ground.[14]

Mexican track workers stood among the poorest Mexican immigrants and built temporary shelters with adobe in the barrio of Chihuahuita in El Paso. The example of one such family in El Paso reveals the harshness of the winters. The El Paso Times reported:

> a family of eighteen persons consisting of the father, mother, their nine children, grandmother and father, and five relatives, was found huddled in a 'dobe [*sic*] house just south of the canal, in the rear of Santa Fe street. All were seated around in the room on the dirt floor, hugging themselves up in their tattered apparel in an effort to keep from freezing to death. There was no fire in the little fireplace and in a box that served as a cupboard were a few but empty dishes.[15]

The dirt floors typically recessed about twelve inches due to the constant sweeping, and when rains came, the sunken floor quickly turned to mud.[16] Railroads failed to develop a policy providing suitable semipermanent housing for their track workers and their families.

Boxcars made convenient temporary housing for both track workers and railroad employers who wanted to hire labor at the lowest possible wage. The cars could be quickly transported wherever needed and easily removed from the tracks to a more permanent location. Even semipermanent boxcar communities, once established, could be quickly and easily relocated whenever railroad officials deemed it necessary. Mexican families often had no choice but to conform to the whims of the railroad bureaucracy and the demands of the economy.[17]

The Santa Fe Railroad developed a Progressive-era employment policy with regard to Mexican immigrant track laborers. Under president E. P. Ripley, the Santa Fe began to implement some of the principles of scientific management outlined by Frederick Winslow Taylor.[18]

To E. P. Ripley, scientific management simply meant the promotion of efficiency and productivity with the least corporate expense. From this perspective, the ideal employee was honest, intelligent, sober, married, church going, hard working, and loyal to the company. Thus in order to retain and reinforce these qualities among their Mexican section hands, the Santa Fe increased the incentive for their Mexican employees to remain with the company rather than seek work elsewhere, or worse, to find comfort in labor unionism. On the other hand, it should be recognized that Mexican workers brought patriarchal relations with them, and most Mexican employees happily sent for their sweethearts, wives, and families. The Santa Fe encouraged young Mexican women and men to relocate on Santa Fe railroad yards in order to reinforce and recreate traditional family values.

This policy reflected the conservative ideology of President Ripley and had its most profound impact on Mexican workers by reinforcing and recreating Mexican patriarchy and ethnic nationalism. An engineer in Kansas summed up this perception well: "The young Mexicans develop into a vagrant class, quite similar to the American hobo... and to obtain a better trained and steadier class of laborers, efforts are made to locate men with families on each section."[19] While the Santa Fe certainly did not house all of its section workers, this housing plan was a widespread attempt to create a stable and permanent work force on the sections. Railroads employed men (section gangs) to maintain particular sections of track as opposed to extra gangs who worked wherever the roadmaster needed them, as in a washout or train wreck. The latter workers irritated foremen when some walked off the job without notice or warning.

Corporate railroad documents are full of references concerning the propensity of single Mexican track workers to simply stop working, walk away, and seek work elsewhere for any number of reasons. The frequency of this mobility is captured well in the following quotation by a southwestern engineer: "As a rule, the men are paid once a month, and this seems sufficient. In most cases they work better when they are entirely without food or money. Whenever they get a dollar or two ahead they seem to want to go somewhere else."[20] John Generella, employment and commissary agent for the Chicago and Northwestern Railway stated: "now the Mexicans have their families on the section, so they will stay."[21] Indeed railroad officials from various lines consistently reasoned that poor housing and substandard meals caused Mexican track workers to quit.[22]

In 1917, a trade journal that specialized in track maintenance published a series of articles on "Mexicans" and "Negroes" as track laborers in response to the great demand for track workers. An engineer on the El Paso & South Western line had this observation: "Mexicans with families are preferred to the single men, as the latter will not stay and work as well."[23]

Santa Fe corporate leadership expected that family reunification would reinforce patriarchal relations in the home. These men had to "support" families; that is, they worked for wages. Railroad officials expected men to work a full schedule, and they expected women to maintain the home and raise children. In so doing, Santa Fe authorities hoped to Americanize their Mexican work force and thereby stem the frequency by which Mexican workers abandoned track work and joined the ranks of the transient or "hobo" population.[24] Santa Fe authorities also believed that accidents in track work could be reduced and efficiency improved if Mexican workers were retained or called back each season. Indeed, they viewed the "happy" wife at home as an essential part of the Safety First and Teamwork campaign.[25]

Implicitly, the railroads and labor recruitment agencies believed that Mexicans were easily pleased with a modicum of housing. A General Manager for the Los Angeles office of the Holmes Supply Company wrote:

> As a rule, the average Mexican laborer and family are satisfied with the most primitive arrangements so far as comfort goes; comfort seems to be a remote consideration with him as compared with kindness and fair treatment. He makes no demand for conveniences, facilities, or arrangements for his well-being, accepting a box car with wooden bunks and no

> windows with the same readiness that he would a modern bunk car, or a new section house fitted up with the latest pattern in steel bunks.[26]

Certainly railroads could expect traquero families to be more complacent, but complacency did not necessarily translate into company loyalty. Even traquero families picked up and left "comfortable quarters to work with or near some remote relative or friend."[27]

Participation in strikes or unionization efforts severely jeopardized family relations because Mexican workers lived on Santa Fe property rent free; once terminated, they risked eviction. Mexican women were clearly important to the Santa Fe in that railroad executives expected women to fulfill their gender-based roles as dependent wives who would pressure their menfolk to avoid involvement in strikes and union activity. Furthermore, women constituted an unpaid labor force whose primary work in the home was to support male employees. Therefore, we might understand Mexican immigrant women's relationship to the railroad company as an economic one. Beyond the biological reproduction of a "common labor" force, mexicanas represented an informal and unpaid labor force and an essential part of the new scientifically managed railroad hierarchy. And for their part, Mexican women appeared no less willing to support their husbands and neighbors during strike activity as has been shown in Chapter 4.

The railroad industry concerned itself with the morality of their track workers. Among the reasons they preferred Mexican workers over African American workers was the willingness of Mexican women to leave Mexico and join their menfolk on American railroads. Apparently African American track workers only reluctantly brought their families along to live under the conditions of itinerant laborers. Certainly most Black workers could obtain better paying jobs in the Midwest and generally avoided track work, but a Mr. Courtney, Chief Clerk of the Illinois Central, erroneously attributed their reluctance to do track work to the difficulty of obtaining women for sexual gratification while at work. He stated:

> They have to take a negro woman along with a colored gang, and if [the gang is] large, a couple [of women]. They [*sic*] negroes won't stay without their sexual satisfaction. Railroads

> [*sic*] officials on inspection have even found them in intercourse in tool sheds, etc.[28]

Apparently the racist idea of the oversexed Black male did not apply to Mexican or track workers, but this stereotype was obviously widespread and probably served to limit the number of Black track workers as well.

Beginning in 1906, the Santa Fe encouraged its Mexican section workers to send for their families, and assisted them with free passes from El Paso to their new homes. The Santa Fe provided Mexican workers and their families with free boxcar housing on Santa Fe property. These were the first Mexican boxcar settlements and in many areas of the West and Southwest, these working-class enclaves became the nucleus of growing Mexican immigrant communities.

Boxcar settlements perhaps symbolize the most significantly visible evidence of Mexican immigrant ties to the railroad industry in the United States and especially in the Southwest. Mexican boxcar settlements could be found as far east as Pittsburgh and Chicago, and as far west as San Diego and Los Angeles.[29] As ubiquitous as these boxcar settlements were, only a few will be discussed here. The areas of Chicago, Topeka, El Paso, and San Bernardino were particularly important hubs of the railroad industry that connected the East with the West and the United States with Mexico. In each of these hub cities, the railroads fanned out in all directions. And among the various transcontinental railroads, the Santa Fe pioneered in the labor policies followed by other lines.

Boxcar housing appeared in two basic forms. The first type provided temporary housing for grading and construction gangs (building new roads) and the second type appeared to be more or less permanent housing for section and extra gangs (maintenance). The first type of housing preceded the second, which were also referred to as rolling villages. Railroads shipped these mobile villages to outlying locations as needed, and the second type of boxcar housing was literally removed from the tracks and usually located on the periphery of railroad company yards specifically reserved for section gangs. Both types were provided only with a wood stove, and families used scrap wood and boxes to substitute for tables, chairs, and beds. Residents in semipermanent boxcar homes eventually bought regular furniture and made basic improvements. More importantly, each family was responsible for meal preparation as opposed to meals prepared by a commissary.

Although the earliest boxcar housing was probably occupied by single Mexican males, this was not the overwhelming experience especially after the turn of the century. In many cases, boxcar housing was located in or adjacent to the roundhouse. It was not until families began to arrive that the railroad relocated boxcars and other forms of housing to the periphery of the yards, but still upon company property.

During the spring and summer months, railroads hired seasonal laborers on extra gangs and occasionally allowed for families to accompany them on the line as extra incentive for Mexican immigrant track workers. As in Texas, these boxcars functioned as rolling villages. In 1902 the Southern Pacific Railroad employed Mexican trackmen in this way. Packed together in approximately thirty boxcars, as many as four families were housed in each car. An observer described the car this way:

> The car where we were entertained, for entertained we were with all the gentleness and hospitality of the Latin race, contained two families—five children and their parents. No sign of partitions, curtains or privacy were in evidence and on the common stove, placed in the open doorway the joint family meal of tortillas and beans were cooking.[30]

The boxcars themselves provided little heat during the winters and virtually no protection during the summers. The following description provides a glimpse of how Mexican women and children coped with the oppressive Los Angeles heat:

> Some cars contain as many as four families, packed together like cattle, and on hot days these ovens are unbearable, forcing little ones with their mothers to turn out and seek shelter underneath amidst the wheels.[31]

Such conditions created a health hazard. Illness was common and so were accidents.

Boxcar housing for extra gangs probably represented the most oppressive form of labor and housing conditions on the railroad. The company store, or commissary referred to as the comisaria, lay at the center of this labor system. For example, the commissary provided food and supplies to the workers and their families on demand with the balance deducted from their monthly pay.

Well into the early twentieth century, railroads employed track workers and housed their families in mobile boxcars and paid them scrip redeemable only at the commissary car. In this way, railroad companies compelled workers to purchase their food, tools, clothing, and other supplies at double the normal cost or more.[32] To be sure, this kind of boxcar arrangement was operated by private employment agencies that specialized in recruiting Mexican workers for specific railroads.

The rolling boxcar villages were operated much the same as bunk cars for solos, hence the necessity of commissary cars.[33] The presence of women and children represented the essential difference from the bunk cars for solos. Most if not all of these track workers accrued serious indebtedness to the employment agency at the end of the work season. This kind of exploitation taught Mexican immigrant workers and their families not to hire on with employment agencies and labor contractors in subsequent seasons. Worker indebtedness frequently went unreconciled, but employment agencies minimized their losses with the profits gained by employment fees and overpriced goods.[34]

The experience of Juana Calderón in Los Angeles illustrates the daily life in these mobile boxcar camps that were most common throughout the 1880s and 1890s but less so into the 1900s. In 1902, Calderón spoke out for approximately 120 women who lived in bare mobile boxcars with their children and husbands who worked as track labor for the Southern Pacific Railroad. Señora Calderón described the conditions of employment to a white woman writer for the *Socialist*:

> Yes, Madame, my husband is at work on the road, we get a dollar a day and have to buy everything for the family out of this and they charge such high prices at the company's store. [35]

Calderón then displayed her monthly yellow bill and explained how the company charged their Mexican track workers for foodstuffs and supplies. One side of the receipt listed different kinds of food and clothes in Spanish and English, and the other side gave the corresponding weights and prices. She further indicated other expenses not listed on the receipt: a monthly fifty-cent hospital fee and a two-dollar deduction for taxes. The comisaria required her husband, José Calderón, to show six continuous months of these itemized bills in order to be eligible for free return passage to El Paso. However, six months' service to the Southern Pacific did not automatically guarantee free passage. Juana Calderón revealed an acute sense of oppressiveness that many traquero families must have also felt: "We cannot save any money, with such high prices and taxes, leaving us always in debt, so we will probably always stay with the [Southern Pacific] Company."[36]

It is impossible to determine the precise extent to which each of the railroads employing Mexican workers (with families) practiced this extreme form of exploitation, but available documents strongly indicate that the practice was widespread. Indeed, this form of oppression likely forced Mexican track workers to live in makeshift camps along the tracks as previously discussed.

The second type of boxcar housing that the railroads provided for Mexican laborers and almost exclusively associated with Mexican community formation consisted of semipermanent settlements. Few of these boxcar communities survived beyond World War II.

With the trucks (wheel assembly) removed, officials typically arranged the boxcars in rows or in courtyard patterns. The number of boxcars in a single community varied from one region to the next, but the available evidence suggests that some boxcar settlements ranged from one or two cars to as many as a dozen or so. Each car contained a single partition, which made two units, and each car was intended to house at least two families. As stated earlier, not all Mexican track workers and their families lived in boxcars. Others lived rent-free on company-owned section houses or privately owned boarding houses.

In 1911, a division engineer in Kansas designed an inexpensive plan for Mexican railroad workers to build their own homes. Other railroads quickly adopted this plan in order to retain their Mexican section forces. Unlike the soils in Arizona and New Mexico, the soil in Kansas did not mix with water and stand up to rain like adobe. Shanties made with mud roofs caved in during the rains, compelling Mexican workers and their families to leave the railroads, thus reducing the availability of section labor. The plan called for making use of scrap materials that could be found lying around the railroad yards. Plan A (see Illustration 5.1.) would be erected in "inconspicuous" locations along the main line and branch lines:

> material being more readily secured, as either sawed or hewn ties will suffice. Side walls are formed by ties set on end 18 in. in the ground, with the cracks chinked with mud or concrete made from rock screenings. Where the cracks are wide they are covered with scrap boards. The partition is built in the same way as the outside walls. Old rails are used for the rafters for the roof, over which a layer of 1 in. sheeting and 2 in. of concrete made from screenings and reinforced with wire netting or second hand barbed wire are placed. One length of 8 in. sewer pipe is used for the chimney. A 3 in. layer of concrete is placed on the floor in some cases, and in others simply an earth floor is provided.[37]

On the other hand, the designer intended for Plan B to be erected at "conspicuous" places along the main line:

> In this case old bridge ties are laid flatwise and shaped with mud or adobe, the partition being made in the same way.[38]

No clear explanation is given as to why the designer intended to locate Plan A houses in "inconspicuous" places along branch lines, and simultaneously locate Plan B houses in "conspicuous" locales along the main line. One possible explanation is that the designer regarded Plan B houses to be more aesthetically pleasing than Plan A houses and therefore less of an embarrassment to the railroad and traveling public. The Santa Fe and other railroads experimented with other forms of housing as well.[39]

ILLUSTRATION 2
SECTION LABORER'S HOUSE FOR THE MAIN LINES

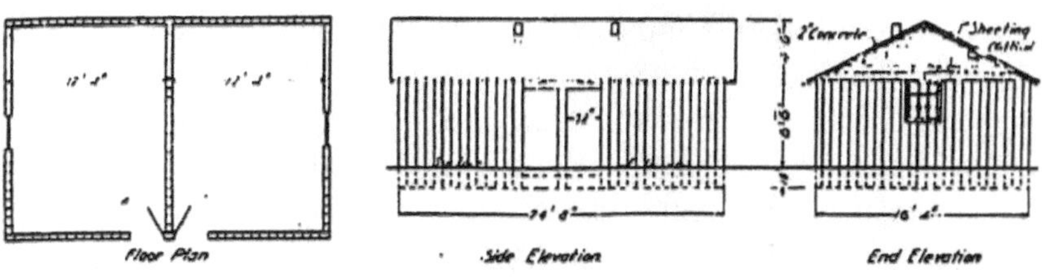

PLAN A - SECTION LABORER'S HOUSE FOR BRANCH LINES

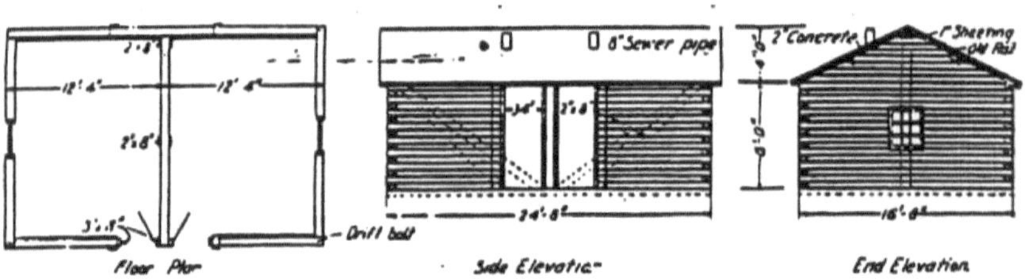

PLAN B - SECTION LABORER'S HOUSE FOR MAIN LINES

Source: *Santa Fe Magazine*, September 1911, 75.

In the Central Plains region the Santa Fe made "hook" houses available to Mexican track workers. Mexican workers constructed these prefabricated "hook" houses themselves. The walls fastened to the roof at the corners and each unit had one door and two windows. The floor consisted of wood, and the entire room measured approximately twenty feet square. Jesús Ramírez recalls that:

> they were like flat boxes, then unfolded and latched together, with what looked like door latches. Later many families added rooms; they built them with railroad ties and patched the cracks with mud to keep out the wind and rain.[40]

In 1921, the Santa Fe built twenty brick and concrete apartments of two rooms each to replace the shacks used by Mexican workers in Newton, Kansas. Seven years later the Santa Fe built approximately twenty-five little brick houses called Las Casitas (Little Houses) by the traqueros in Emporia, Kansas. However, these houses had no partition inside the dwellings, and monthly rent cost the workers fifty cents per room. And like the boxcars, they had no electricity and a wood stove provided the only source of heat. However, the Santa Fe piped in water for the families living there; there was one faucet for eight families.[41]

Fidel Ybarra, at age sixty-three had worked forty-four years as a section hand for the Santa Fe Railway and lived in Las Casitas during the 1920s as a small child. He recalls:

> Those houses at Gladstone were made of one-by-twelves, and the gaps between the boards were covered with thin

> wood strips. Tarpaper over the roof. No insulation except my mother's wallpaper. We had two rooms, about fourteen by twenty—that was the whole house. After the kids come [*sic*] along there was twelve of us in there. No electricity, no running water. There was six other chanties [*sic*], some just one room, and all we had was one outhouse with two doors, a men's and women's, and just one seat in them: thirty people and two seats. We had a pump for water. But Santa Fe didn't charge us nothing [*sic*] to live there, and in the winter the company sent in a car of old track ties and pieces of depots and boxcars, and we'd unload it and chop the wood up for our stove, but the place was still cold.[42]

Teresa Ybarra, Fidel's wife, quickly added:

> Oh yeahhh [*sic*]. If we went to Emporia to visit overnight, when we came home everything was frozen.[43]

Domingo López of Topeka recalls that mexicanos there affectionately referred to the boxcar community as La Yarda (The Yard). During the 1910s, La Yarda represented the center of the Mexican railroad worker community life and also functioned as a social springboard for Mexican immigrant acculturation in Topeka.[44]

Boxcar communities in Chicago shared similarities with La Yarda. Each unit housed the track worker, his wife, children, and sometimes extended family. The Neria family provided a useful glimpse into the lives of at least one Mexican track worker household. Between 1916 and 1928, the Neria family lived in various boxcar communities before settling in Chicago. With the assistance of his mother, Mr. Neria provided for four children and sent them to local schools. A researcher in July of 1928 observed that the elder matriarch was "the one of authority and seemed to be a woman of some character."[45]

Boxcar housing amenities varied from one railroad to another. Some companies equipped boxcars with new wood burning stoves, some had old ones. Some had windows and glass, others had none. In the early 1900s, according to a government report, the Santa Fe and the Southern Pacific lines took "special care to adopt boxcars for housing these men, clean, well ventilated and superior to those furnished on most other roads."[46] Railroads did not always provide sanitation services, filling holes and digging new ones, in a timely manner. But, beyond the services provided by the company, individual families could provide a modicum of comfort.

Some families could afford to buy furniture, others bought or inherited used furnishings from neighbors, and still others used a combination of manufactured furniture and crude boxes and scrap wood for shelves.[47] However most cars came equipped with a wood-burning stove and stove pipe, at least one door, and in most cases a solid wood floor. While most boxcars did not come equipped with modern plumbing, inhabitants drew water from a communal pump provided at no charge by the railroads. Families shared and maintained toilet facilities, most often at least two outhouses, one for women and one for men.[48]

The families with the most elaborate furnishings and largest boxcar homes with room additions tended to be those who lived in a single boxcar home for a few years. Recent arrivals could generally expect to inhabit the least attractive boxcars and possess virtually no furniture. Solos typically inhabited a single boxcar which was often crowded with approximately fifteen men. This posed a health risk both to themselves and their neighbors, especially due to smallpox, influenza, and tuberculosis. Due to the itinerant nature of track work, boxcar housing occupied by solos probably appeared less attractive since the men had less time to plant gardens, clear shrubbery, furnish indoors, and do general household maintenance.

In 1928, Paul S. Taylor interviewed and observed the living conditions of numerous Mexican track workers and their families. The oldest resident of the Manheim Camp in northwest Chicago had been there since 1920. This traquero along with his extended family, lived in what might have been the most impressive boxcar home in the Midwest. According to Taylor:

> It had expanded from a single dwelling or boxcar to the addition of three rooms and a good covered porch with a good floor. He had a mess of about thirty chickens and over one hundred chicks with a good chickenhouse. He had a large garden of tomatoes, potatoes and some other vegetable [*sic*] and was quite proud of it.[49]

One distinguishing difference between the Manheim Camp and most other boxcar camps is that the camp was located about a mile away from the redonda (roundhouse) and main yards of the railroad.

The Manheim residents preferred living away from the yards for a variety of reasons. First, Mexican families preferred this camp, located on the outskirts of town, for the safety it afforded their children and others. The children could play more freely in the surrounding countryside without worry of danger around the yards. They escaped the smoke and soot from the roundhouse which made the air dirty and difficult to breathe. Traquero families dreaded the soot-ruined laundry on the clothesline.[50] Indeed, the soot fell on everything and constantly reminded them of the industrial nature of their housing.

Taylor further noted that this family decorated the inside of the boxcar walls with attractive wallpaper, and the car itself appeared neatly organized and clean. And like the inhabitants of other Mexican boxcar homes, the Nerias adorned their walls with photos and religious decorations from Mexico.[51] While most families appeared to be left unmolested by railroad officials while they lived in these boxcars, in the 1920s some railroads began to require maintenance-of-way foremen to inspect the camps. Some foremen chauvinistically took credit for teaching Mexican women simple laundry tasks. R. W. Nicol, charged with inspecting Mexican quarters on the Elgin, Joliet & Eastern Railroad, reported: "I have seen Mexican women's dresses carefully hung up on hangers and covered with cloth to keep them clean. They were taught by the foremen."[52] Elsewhere, Nicol reported that the EJ & E Railroad hired a "matron" to teach mexicanas how to:

> prepare food which is more nourishing and suitable in the North than the inadequate diet on which they now try to live. I recently watched one of our Mexicans eating lunch which had been prepared at home. It consisted of flour tortillas, a sort of pancake made of flour, salt and water. Surely such food has insufficient nourishment for a man engaged in track work.[53]

The presence of certain features in boxcar communities often indicated long or short-term residence. For example, the presence of poultry, gardens, and pets could be found in the boxcar communities of long-time residents. Indeed, gardens provided families with their herbs and spices, otherwise unavailable in markets in the Midwest. Along with fresh eggs, vegetable gardens allowed Mexican families to stretch their income and supplement their diets with wholesome foods. A few families even raised pigs and goats.[54] Beyond the ability to stretch their meager funds, gardening and animal husbandry symbolized traquero social ties to rural life in Mexico. Indeed, men and women took silent pride in growing among other things, tomatoes, onions, cilantro, and chiles. Many if not most itinerant boxcar residents also took tremendous pride in their flower gardens and plants. One friendly observer noted: "The humblest section house or slum *jacal* (thatched hut) built of old piano boxes, has its flowers or potted plants."[55]

The physical presence of the huge rectangular boxcars, so ubiquitous and immediately recognizable, appeared to be the most common feature of boxcar settlements. Indeed, for many Mexican railroad workers in the United States, the physical atmosphere provided by boxcar life paralleled their work and living conditions for those workers who had previously been employed by the *Ferrocarril Naciónal de Mexico* (Mexican National Railway). Indeed, for many traqueros, living in boxcars was not a new experience. Another common feature of boxcar life was a stockpile of old railroad ties and a saw horse. Since each family required fuel for their stoves, men, women and children routinely chopped wood for heat and cooking. The abundance of old railroad ties provided a good source of fuel and was less expensive than coal. In order to make efficient use of each tie from the wood-pile, traqueros cut the wood in a very specific manner. According to José Cuevas of Topeka, he cut each tie with a handsaw into twelve-inch lengths. After he cut the desired number of ties, he stood the twelve-inch lengths on end and split them into four pieces with an axe. These lengths fit easily into the stove.[56]

Burning railroad ties for fuel also posed potential health hazards. Creosote and other chemicals used to preserve railroad ties created noxious smoke when burned. Boxcar residents had to be careful to keep the stovepipe clear of obstructions and in proper condition. Indeed, boxcars, shanties, or tie shacks frequently burned down as a result of accidents or improperly operated wood stoves.[57]

Railroads typically located boxcar communities on or near the railroad yards in order to have immediate access to section laborers. For instance, some of the boxcar communities in Southern California were *El Hoyo* (The Hole), *La Sección de la Calle Tres* (The Third Street Section), *Los Flats* and *La Reserva* (The Reserve).[58] In El Paso, Texas, the track worker community was known as *chihuahuita*, since many of its residents came from Chihuahua.[59] In *Kianses* (Kansas) some of the Mexican track worker colonies were *La Yarda*, *Los Bottoms*, *Las Casitas* (The Small Houses), *Los Ranchitos* (The Small Ranches),[60] *El Huarache* (The Sandal), and *La Quemada* (The Burned) to name a few.[61] In Michigan, a large number of Mexican families lived near the yards on the west end of Detroit, near Sixth Avenue and Plum Street as well as in the yards on Michigan Avenue near the depot.[62] According to Taylor, the Mexican community in Chicago was concentrated in the "colored sections where the poorer of all nationalities live" on or near Black and Pennsylvania Avenues.[63] The boxcar communities of Silvis, Rock Island, Moline and East Moline, Illinois, were located next to one another along the Mississippi River. Perhaps the largest boxcar and railroad worker community was in South Chicago.[64]

While Mexican railroad workers and their families had no choice but to live in boxcar communities, Euro-Americans considered such communities a health hazard and a nuisance. The white community in San Bernardino, for example, referred to the traquero camp as "Cholo Hollow." The larger Mexican immigrant community on the west side of town surrounded this camp. Over half of the city's Mexican population lived in the west side colony, and though most of the Mexican families lived outside of this camp, their presence in the yards was a constant source of annoyance to the city authorities.

In San Bernardino and elsewhere, newspapers and whites commonly referred to the Mexican part of town as "cholotown," "choloville," "Greaser town," "Sonoratown," or "Little Mexico" and usually associated crime, drunkenness, drug use, and other forms of disreputable behavior to the Mexican side of town. Upon witnessing the arrival of a gang of traqueros, a prominent judge in San Bernardino, California, remarked: "A criminal class all of them."[65] Depending on the circumstances, city officials and local civic associations attempted to make distinctions between old California Mexicans and "cholos." The former "have the respect of all their neighbers [*sic*]."[66]

City officials regularly complained to the Santa Fe about the unsightly dwellings. According to one city official, Cholo Hollow was a source of embarrassment to tourists and travelers passing through San Bernardino on the passenger trains. In 1909, city and Santa Fe officials decided to relocate the Mexican boxcar community. A local newspaper described the reason for the proposed relocation:

> The present location of the cholo quarters, in the wye [*sic*] at the west end of the yards, is in full view of passengers on all

> trains leaving or arriving on not only the line to and from Los Angeles but Orange [County] as well, and in consequence is passed by every train through this city. For years the camp has been an eye-sore, marring the entrance to the city.[67]

Economic necessity compelled Mexican track workers and their families in San Bernardino to share their dwellings. And as the following quote shows, local authorities mistakenly interpreted cramped quarters as a genuine cultural trait of all working-class Mexicans. one observer sarcastically noted that railroad workers and their families in Cholo Hollow "packed [themselves] together like sardines in a box, as is the custom among Peons."[68] The observations of an El Paso & South Western engineer from Douglas, Arizona, seems to concur with the previous description. He noted that Mexican track worker "houses should have rooms about 12 by 16 ft., and in these rooms one and sometimes two families will live, or from six to eight single men."[69]

By 1928, the "world's largest interurban system," the Pacific Electric Company in Los Angeles, built twenty-two "model" labor camps for its Mexican workers. Each labor camp housed hundreds of workers. The Pacific Electric served four major counties in Southern California and employed hundreds of Mexican track hands and general common laborers. The company hired two men at each camp to maintain the grounds and also provided a trained nurse to see to the health needs of the employees and their families.[70] A writer for a trade magazine boasted of the Pacific Electric's handling of the so-called Mexican problem:

> These camps, while they may not be the highest type of modern living quarters, are at any rate clean, sanitary and wholesome; they are fulfilling an important economic need and are relieving charitable organizations of considerable humanitarian work that would otherwise be necessary. Above all, it is felt that they are meeting a problem of international relations in a manner creditable to the company and the community which it serves. The company feels that it is being well repaid for the effort put forth and the work is to be continued.[71]

By the 1920s, railroads continued to house Mexican track workers in boxcars, but as suggested by this quote, some railroads and interurban lines had seriously begun to implement official housing policies for Mexican labor. However, the racism and segregation continued largely unabated. In the mid-1920s the cover page of the *Pacific Electric Magazine*, drew the reader's attention to the "Spic-spanness" of its Mexican section camps.[72]

Although boxcar communities did not represent the first centers of Mexican community formation in the United States, these incipient communities marked the beginning of some Mexican immigrant settlements. The centrality of the boxcar community to Mexican immigrant life in the United States is due largely to its proximity to the railroad yards. But the convenience of proximity and free rent took a terrible toll on the inhabitants.

Accidents were a common occurrence throughout virtually all Mexican—and non-Mexican—railroad worker communities in the United States. A few examples of these accidents provide a cold glimpse into the life of Mexican boxcar communities. In early 1904, a Southern Pacific train bound for Santa Barbara from Los Angeles struck an unknown elderly mexicano. The train struck him as he walked alongside the track and knocked him off a bridge. The Southern Pacific authorities reported that:

> The train was running about thirty miles an hour. The engine passed him, but the cab steps struck him on the shoulder, hurling him to the ground twenty feet below. In falling the man's head struck the cement abutments; causing concussion of the brain. He died without regaining consciousness.[73]

In November 1905, Matias Romero of El Paso apparently stole a ride on an incoming train only to face tragedy. A local newspaper reported:

> It is thought that Romero was riding the bumpers between two cars and was climbing down to alight as he missed his footing and fell on the track, flat of [*sic*] his back. The train was going at the rate of about ten miles an hour and it was not an instant before his body was caught and ground to pieces. The trunk was separated from the legs and fully twelve cars passed over the body.[74]

Children could be injured killed while playing near or in the railroad yards and tracks. Twelve-year-old Eustacio Carrasco of El Paso attempted to jump off a moving freight train when he slipped and fell. He reportedly lost his grip and "fell under one of the cars, the wheels of which passed over both legs below the knee, cutting them off as if they had been reeds. Some part of the car smashed the boy's head which was smashed in."[75] Similarly, eight-year-old Elias Martinez apparently tried to steal a ride on a trailer car and when he jumped off somehow became caught under the wheels. The wheels crushed his foot, and although he survived, doctors amputated his mangled foot above the ankle.[76]

Aside from the dangers of living near railroad tracks, streetcars represented another form of danger, particularly to children playing in the streets. In the winter of 1909, a streetcar in El Paso hit and dragged eleven-year-old Refugio Armenta several yards as he played with his kite. The impact crushed his head and the front wheels passed over both his left hand and foot. A local paper described the scene:

> Meantime scenes probably as heartrending as any that have ever been enacted, drew a morbid crowd. Three or four hundred persons gathered around. When the little fellow's mother was notified she became frantic with grief. Friends took her to her home at 1220 South Stanton, while the little lad's father, shaken with sobs, knelt heart broken, beside the crushed form of his little boy.[77]

Carmen Rangel González recalls the sounds of the trains as a child behind her home in Winslow, Arizona. Her experience one afternoon reveals still another perspective of the danger and potential danger of work and life near the railroads. She said:

> I grew up hearing the tooting of trains in our backyard. This was common to my childhood, just as the ice cream truck was for my kids. The desert w(h]ere we lived was beautiful and often very hot. The highlight of my day was to watch the trains speed by. Other Mexican children of the same housing complex would race to the fence and climb to the top just in time to wave to the engineer as he passed. Each time we heard a train, it was a race to see who could get the first wave. One evening around dusk, we heard a tooting train as usual, but this time it sounded funny—quick, panicked toots. In routine

> we ran to the fence, and saw a man pumping a hand car violently towards us over a short bridge, not knowing there was train on his tail. My heart fell and I screamed out, "Papa!" Just as my mother came running to the fence, Papá passed over the bridge and dove off the tracks pulling the cart with him. I thought he was dead, but we ran to find only bruised and scratched at the bottom of a small ditch.[78]

Although railroads provided a plentiful supply of firewood and water, the case of 76-year-old Demesia García de Martinez reveals how dangerous it was to merely gather fuel to heat the stove. In February 1907, Santa Fe yard authorities in El Paso theorized that Mrs. Martinez:

> had crawled under the car either in search of wood or small pieces of coal and that when a switch engine picked up the cars, the life was crushed out of her before she could escape. The left leg was crushed from the hip to the shoetop and the right leg from just above the knee to the ankle. The middle and the ring finger of the left hand were broken.[79]

The case of an elderly mexicano gathering pieces of coal along the tracks also exemplifies the danger of section house life in El Paso. In 1905, a train severed the legs of E. Cervantes near the hips. Cervantes had apparently crawled under a string of cars and tried to escape when a Santa Fe switch engine set the cars in motion. Railroad authorities took him to the county hospital where doctors decided that they could do nothing to save him. Cervantes died soon after. An inquest followed his death and exonerated the trainmen.[80]

Streetcar accidents occurred frequently throughout the country's major cities, including El Paso. In 1905, a streetcar struck an unnamed mexicana and seriously injured her as she walked across the tracks. Hard of hearing, and dressed in a traditional black rebozo, she did not hear or see the car when it struck her from behind. The local press reported:

> The front part of the car struck the old woman on the side of the head, inflicting an ugly scalp wound, which rendered her unconscious and caused a profuse flow of blood, making it appear for a time that she was fatally injured.[81]

While this woman survived her head wound, the injury must have been especially traumatic for someone her age. Beyond these ghastly cases, such experiences reveal that life near the railroad tracks was dangerous for all sectors of the Mexican working-class community, including children, women, and the elderly.

Clearly, Mexican workers did not create these hazardous conditions associated with boxcar communities. These conditions existed throughout the country wherever immigrant groups worked and lived among the roundhouses and railroad yards. But occasionally, traquero families sought out legal assistance in order to pursue justice and redress in the courts. In 1902, a lawsuit for $13,750 was filed against the El Paso and Northeastern Railroad for failing to signal at an intersection where a twelve-year-old boy was hit by a train and lost both his hands. In most cases, Mexican families had no legal recourse and were merely left to cope with their grief in private. Active mutualistas provided some insurance to its members, but the assistance rarely amounted to a substantial amount. In 1922 the Sociedad Ignacio Zaragosa in Chicago provided $100 for a lost arm, foot, or eye. This sociedad also paid the injured member $10 a week for six months.[82] For most injured traqueros and their families, American laws and customs remained unfamiliar, and combined with linguistic differences, few if any ever successfully sued railroad companies for monetary awards. By the late 1920s and 1930s this had begun to change due to the efforts of a few Mexican American lawyers.[83]

Diseases also plagued Mexican track worker communities, as they did virtually all immigrant communities during this time. Smallpox, influenza, measles, among other diseases commonly visited track worker camps, especially itinerant camps. In El Paso's Chihuahuita, a Mexican immigrant enclave, disease reached epidemic proportions. In 1883 local officials expressed alarm at "unsightly" jacales as possible breeders of disease. In 1892 diphtheria surfaced among the Mexican population, and in 1904 it recurred in epidemic proportions. El Paso officials deported suspected diseased Mexicans to Ciudad Juarez regardless of citizenship.[84]

Other agencies approached disease among Mexican immigrants more humanely. According to the *Survey*, the "chief difficulty is to prevent overcrowding, to seek out dark and unventilated rooms, and to keep up a constant search for contagious diseases." Between 1907 and 1911, tuberculosis caused 1,428 deaths in El Paso, including 457 Mexicans.[85]

One radical white woman writing for the *Socialist* had this critical observation on the causes of squalor and disease among the Mexican camps. She said:

> Not only small pox [*sic*], but any other contagious disease that springs from foul surroundings and impropper [*sic*] sanitation may start in these camps. One glance at the ground about the cars reveals filth of every description and lack of sinks shows the cause—for all of which the railroad is responsible.[86]

Indeed, the horrible conditions of working-class immigrant groups drew the attention of other muckraking journalists during the Progressive Era.

While railroads segregated Mexicans on the job, in schools, in housing, and in public services, traqueros and their families—like other immigrant and non-white workers—created a certain degree of comfort for themselves in their own segregated communities. Boxcar communities and other railroad-owned housing served several important functions. Some Mexican families lived "rent free" and saved money to pay off debts, help family in Mexico, and make major purchases. The proximity of boxcar housing to the yards proved to be a mixed blessing. The convenience of living close to other Mexican families enabled traquero families to assist one another with child care, family illness, foodstuffs, meals, and small loans. Perhaps most importantly, living in boxcar communities allowed Mexican railroad workers and their families to retain and reinforce their language and cultural identity.

Notwithstanding the positive sense of ethnic solidarity that barrioization created and reinforced, the sting of discrimination is recalled bitterly by most traqueros and their families. Fidel Ybarra recalls that in "Cottonwood or Emporia, Topeka too, we couldn't get served in restaurants, but at some places you could take your food out or go around to the back to the kitchen. We couldn't get no haircuts neither [*sic*]."[87] Irate Colton city officials limited "cholos" to receive their mail on Sundays at the rear of the post office.[88]

Mexican workers naturally emphasized cultural boundary maintenance especially in areas where Mexicans faced social isolation and alienation. In the Plains and Midwest where no earlier Mexican barrios existed, the arrival of traqueros marked the beginning of one form of barrioization. This cultural isolation gave rise to what we might understand as "temporary nationalism."[89] This ethnic nationalism resulted from the railroad segregation of Mexicans on the job as well as Mexican workers reinforcing Spanish and other Mexican customs among themselves.

Living in close proximity to other Mexican families gave rise to a common set of experiences in the United States and therefore helped reinforce Mexican cultural traditions. This also served as a filter for American culture and values. Boxcar communities functioned as way stations for newly arrived Mexican immigrant families. New immigrants could therefore cushion their adjustment to American customs and society through the familiar institutions of the family and the barrio.

The reality of Mexican track worker community and boxcar life combined opportunities and hardships. Few Mexican railroad workers ever seriously challenged the basic structure of racial and economic inequality they faced. Mexicans took solace in their private lives and reckoned that they would eventually return to Mexico when it was safe to do so. Still, hundreds of thousands of Mexican immigrants ultimately remained in the United States. La patria chica (the homeland) remained the reference point for the majority of the Immigrant Generation.[90] Like other immigrants, their children would have a different outlook and a different political identity. How traqueros interpreted and coped with the social and economic conditions in the United States reveals the centrality of the railroad to the experience of working-class Mexicans in the United States during the early twentieth century.

The principles of scientific management as implemented by the Santa Fe railroad and perhaps other railroads certainly affected the recruitment and employment and housing patterns of Mexican railroad workers, but far more important were the Mexican women and men who made the daily choices about the conduct and quality of their lives, albeit in a context shaped by the harsh realities of railroad work in the United States and Mexico. Unfortunately for Mexican workers, the railroad industry was in decline. By the 1920s automobiles and trucks had captured much of the transportation industry. The skills Mexican workers acquired and the adaptation to the North American work ethic, however, helped them to survive beyond this period and created a tradition of working-class culture in most industrial sectors of the Southwest, Central Plains and Midwest.

chapter 6

Traquero culture

THIS CHAPTER EXAMINES CULTURAL RELATIONSHIPS AMONG Mexican railroad workers and their families both at home and on the job. Furthermore, it shows how a Mexican working-class culture evolved to become a distinct railroad-worker culture inextricably tied to work on the railroad, especially track work.[1]

While the experiences and behaviors of Mexican railroad workers and their families were not uniform, certain cultural aspects such as adaptability and resiliency characterized Mexican working-class culture. Indeed, cultural continuity and change were mutually inclusive processes. Hispanos and Mexican immigrants adjusted themselves to the new conditions of industrial life. Moreover, their contact with Euro-American institutions—especially schools—slowly transformed Hispanos and Mexicans into what I argue was Mexican railroad-worker culture or traquero culture. Traqueros themselves gave shape and meaning to their lives on a daily basis. With picks, shovels, frying pans, and diapers, traqueros (both men and women) built their lives. Along with the thousands of miles of track that they laid and repaired, they also constructed their own world and made it their own. Cultural change came about largely because the Hispano and Mexican immigrants did not control the formal institutions (jobs, schools, and churches) that transformed their culture. Cultural continuity, on the other hand, operated in the everyday working-class cultural experiences and were evident in the family, religious, recreational activities, and voluntary associations.

The penetration of the railroads into Indian lands, former Mexican lands, and Hispano lands in Colorado and New Mexico fundamentally changed the local and regional economy from a pre-capitalist one to a rapidly expanding industrial capitalist economy. For example, the railroad in many instances displaced American Indian men from their villages for the promise of jobs and cash wages. In return, groups like the Hopi, Navajo, and Apache were compelled to give up large portions of their traditional lands to the railroads. Many American Indian track workers never returned to their villages and homes thereby further contributing to the dislocation of American Indian villages. Indeed, the destruction of indigenous cultures was one of the essential purposes of the railroad according to the federal government.[2] The railroad allowed for the rapid troop deployment against Indian resistance in the West and facilitated forced relocation of American Indians from their traditional lands.

During the two decades following the completion of the first transcontinental railroad in 1869, the major railroads in the West hired any and all immigrant workers, including non-immigrants like American Indians and Hispanos, to build the railroads.[3] In the West, the railroads employed large numbers of Chinese and Japanese in track work, and in the East and Midwest, railroads employed Irish, Greeks, Poles, Italians, Slovaks, and other Eastern European immigrants. And as shown in chapter 2, the employment of immigrant Chinese track workers was curtailed by the Chinese Exclusion Act of 1882. In addition, Japanese workers, who were not as widely employed by the railroad, found other employment in agriculture or clustered in urban centers like San Francisco. European immigrant workers also typically found better paying employment in mining, manufacturing, and factories as laborers and semi-skilled workers and created a vacuum in track work that remained chronically unfilled.

Because the vast majority of Hispanos were land-poor farmers or landless workers, they controlled few large formal institutions such as industries, schools, or government. However, they did control informal institutions such as the family that allowed them to survive as an ethnic community. The encroachment of the railroad placed direct pressure on Hispano and Indian families first, and then later upon Mexican immigrant families. Despite these pressures, the family was perhaps the basic social unit and cultural institution that allowed Hispano and Mexican families to resist forced Americanization. Within this social unit, women, more than men, were central to the maintenance of cultural traditions. Although no large-scale research has been done on late nineteenth-century and early twentieth-century

Hispano and Mexican peasant and working-class families, it appears that the family formed an important social and economic unit.

With the completion of the first transcontinental railroad in 1869 and the boom in railroad construction in the West and Southwest from that point forward, Mexican immigrant workers and their families began to trickle into the Southwest and Central Plains. Unlike those of Mexican background already living in the Southwest, the immigrants brought their own set of cultural values and assumptions which they transferred to unsettled areas of the Southwest, Central Plains, and Midwest. Thus the movement of Mexican men and women into the United States not only represented a labor flow, but a cultural migration during the early twentieth century.

However, cultural differences between Mexican immigrants and native-born Mexicans soon manifested themselves. One striking example of this difference concerned the Catholic Church. As discussed in the previous chapter, the difference was that Hispano or Mexican American Catholicism was wedded to the Southwest since 1598 while Mexican railroad workers participated in more Americanized churches. Moreover, the relationship between Mexican parishioners and priests differed geographically. For instance, the Catholic Church in New Mexico tended to reinforce Hispano ethnic identity through Spanish language services, whereas services for Mexican immigrants, such as in the Central Plains and the Midwest, tended to emphasize assimilation by offering services in English. [4] As Mexican communities, especially in the Plains and Midwest, grew in size, some founded their own national churches and specifically recruited Mexican or Mexican American priests to head their parish. Thus, the Catholic Church, depending on the region, served both to maintain Mexican cultural boundaries and to encourage acculturation to American society.

Another significant informal social institution that survived the colonial period in the Southwest was *compadrazgo* (godparenthood). Compadrazgo was an important social network between families that bound them to one another through the ceremony of baptism. Usually one couple, close friends or family, agreed to baptize the child of another upon request. In the case of mortal illness and death, the surviving set of compadres would theoretically assume responsibility for their *ahijados* (godchildren) and raise them to be good Catholics. More importantly, the relationship between compadres was to be held in high regard and often equal to that of blood relatives.

The relationship amounted to a social network between families and close friends that facilitated the sharing of food, clothes, water, chores, and other resources especially during time of crisis. Compadrazgo was central to the culture that Mexican immigrant/migrant workers carried with them. Indeed, compadrazgo was the social mechanism that facilitated chain migration, housing and employment opportunities for Mexican railroad workers as shown in chapter 2. [5] One observer had this telling, albeit not entirely accurate, understanding of compadrazgo, "[Mexican] Tramps are unknown; there is always a relative or compadre who will help out in time of trouble."[6] A Santa Fe official had a similar observation of immigrant Mexican workers. He noted that "They invariably travel in pairs, trios or groups, consisting of relatives, neighbors or compadres."[7] Chain migration and chain occupation links were not peculiar to Mexicans. In his fine study of railroad workers primarily in the Northeast, Walter Licht informs us that the "Family connection represented probably the most important asset in securing employment for all grades of railway workers."[8]

The rise of industrialism in the Southwest also challenged the traditional patriarchal organization of the family. Hispano and Mexican men were often compelled to leave home seasonally to earn wages on section gangs and in so doing left women in charge of domestic responsibilities and those responsibilities previously performed by men. Beyond this gender division of labor for women which included child rearing, food preparation, and other work inside the home, many Mexican women forged their own strategies for survival. It is difficult if not impossible to know exactly how many women were economically dependent on men since census workers typically did a poor job of counting Hispanos. For example, Deena González shows that many Spanish-surnamed women in Santa Fe, New Mexico, who were otherwise considered "unmarried," were actually only left alone temporarily by their husbands who had found seasonal work on the railroads.[9] In 1880 the census enumerated 183 female heads of households in Santa Fe, or 13 percent of the population. González astutely points out that until 1880, relationships between household members were not cited and therefore the assumption "that a man and a woman of similar age were married might be erroneous, especially in a community where households included extended families."[10]

The arrival of the railroad disturbed and in some cases destroyed Hispano family and village life beginning as early as the 1870s. One historian sees the impact of this disruption in seemingly contradictory ways. During the early years of the railroad building in Colorado and New Mexico, the railroads cut across traditional Indian and Hispano communal lands. While Indians and Hispanos did not live among each other without periods of conflict and racial/ethnic tension, nevertheless the Indians and the Hispanos were thrown together in track gangs. According to Deutsch, during the earlier years railroad construction provided increased opportunities for Hispanos and Indians to supplement their personal and family income. These track workers then used their wages to buy more land or to improve the land and crops that they already owned. On the other hand, the railroads themselves often uprooted Hispano villagers from their homes and compelled them to abandon their villages and agrarian livelihood permanently.

González shows that prior to 1880s, a few women like Gertrudis Barceló were astute entrepreneurs who, through her many business dealings and contacts, helped to create a tradition of biculturalism in Santa Fe during the turbulent mid-nineteenth century.[11] However, Barceló was only one of a relatively small number of mexicanas who possessed the wealth and resources to influence relations between Anglos and Mexicans. At the time she died, she owned three houses, a few mules, and tidy savings. [12]

While Mexican women were rarely employed by the railroads to do track work, mexicanas did make use of what resources they had to supplement their income. As discussed in chapter 3, Mexican women created an "informal economy." These women supplemented the family income by taking on unregulated non-wage work; they took in laundry, sewed, and prepared and sold Mexican food at train depots and downtown. Some sold small Mexican curios door-to-door, and hired themselves out for domestic work in white middle-class homes.

Antonia López of Topeka recalls that she and other Mexican girls, as well as many African American girls, would take the trolley to the white parts of town to do house-cleaning, laundry, and other domestic work for white women. The most wealthy of these families were government and Santa Fe railroad officials. Antonia López and women like her could not help but be struck occasionally by the realization that these wealthy white women welcomed them into their homes only as servants.[13]

Perhaps most importantly, the mexicanas' informal economy provided a greater sense of autonomy for women and girls who wanted to supplement the family income. In so doing, Mexican women created their own space in which men were not the sole breadwinners. Consequently, they temporarily softened, to varying degrees, some of the most extreme expressions of patriarchy. The demand for Mexican women wage workers, especially in laundries, packing houses, and garment industries may have had the effect of reducing the degree of patriarchal control. This is not to say, however, that male domination was eliminated. The general expectation from both the Mexican families and railroad companies was that women's work outside the home was only a temporary condition. Although railroad companies were among the first to employ white women exclusively in clerical jobs, the corporate leadership opposed egalitarian relations between their male employees and their wives.[14] As discussed in the previous chapter, the railroad saw its business interest being best served when its employees lived on railroad company property with their families and with patriarchy intact.

Another informal cultural institution that spread with the extension of the railroad was the preparation of Mexican food by both Mexicans and non-Mexicans. Hispano and Mexican railroad workers and their families shared a common mestizo heritage and thereby shared many preferences for meals. At the same time, Mexican railroad workers, like other workers, had different tastes, regional preferences, and different methods of meal preparation. For example, corn tortillas were a staple, but beef and pork were not always available. And while some Mexican workers from northern Mexico preferred beef, beans, and rice, some Hispanos in New Mexico and Colorado preferred mutton and corn. While Hispanos, Pueblo Indians and Mexican immigrants shared some of the same diet preferences, their most common food was *chiles* prepared in a multitude of ways.

Given the heterogeneity of Mexican families on the railroads, it is difficult to arrive at a completely accurate pattern of family changes between 1880 and 1930. As stated earlier, Hispano men were often compelled to leave the home seasonally to do track work or field work to supplement their income; it is probably safe to say that Hispano males were considered the main breadwinners, even though they probably did not work for wages year-round. Mexican immigrant men who found work on the railroads and who eventually brought their wives and families with them to the United States during the late nineteenth and early twentieth century were also considered the main breadwinners. And though their wives and daughters may have taken in laundry or other domestic work for neighbors, the traditionally patriarchal Mexican father and husbands ideally preferred to minimize their wives' work outside the home. Mexican immigrant women found themselves bound to the home for the purposes of cultural maintenance, thereby forming the basic unit of working-class Mexican patriarchy and ethnic community in the United States. Indeed, these Anglos also reinforced this traditional image of Mexicans.

Anglo perception of Mexicans and things Mexicans were mixed. Some hated Mexicans and all things Mexican and others genuinely appreciated Mexican culture. Between these two extremes were all combinations, and perceptions varied regionally. Nonetheless, it is safe to say that most whites felt more antipathy than sympathy toward Mexicans and things Mexican. In 1882, one traveler had the following rather sympathetic observation of the esthetic qualities of tamales:

> a triumph of the Mexican cuisine. It is made of corn meal paste, enclosing a bit of chicken, tied up in a corn husk and boiled. You revolt at it! Let me tell you it is a royal tid-bit and the race that invented tomales [*sic*] can never lose its liberties. [15]

While the preparation and consumption of ethnic foods rarely guaranteed liberties, the fact that patriarchal culture tied women to the kitchen certainly guaranteed some level of freedom for men.

Chorizo, like tamales, required special preparation that was as much a social process as it was an economic enterprise. In areas outside the Southwest where there was not a large and established Mexican community and market place, foods like chorizo, a Mexican sausage, were not readily available and were therefore highly prized. José Cuevas, a worker in the Santa Fe shops at Topeka, recalls that his mother, Celia Cuevas, would enlist the help of her entire household and nearby comadres to prepare vats of chorizo. The process would take a few days and required balancing the labor with child care. Meal preparation continued as usual, albeit in restricted quarters: Celia Cuevas placed large tubs on two benches and filled them with ground meat, chili powder and spices, then mixed it with a large stick. She then covered the tubs with a white cloth and let them sit overnight. In the morning Celia had her children help her prepare the chorizo in bolitas. She gave the *tripas* (intestines bought from the

Hormel packing company) to the children to be given to neighbors or sold. Jose Cuevas remembered that:

> Us kid's job was to take those intestines and blow on 'em. To open 'em up. And my two sisters, my two older sisters would be there with a funnel stuck in that intestine and just crammin' that chorizo in there. Just crammin' it, crammin' it, crammin' it. They'd do that for hours and hours.[16]

Mexican women purchased *tripas* (intestines) and *menudo* (tripe) inexpensively from the meat packing houses at this time since the packers had no use for the intestines. Once they completed the process and the portions were divided up, they sold the remainder at the local market or directly to neighbors. The income derived this way supplemented the income of the male breadwinner and allowed women like Celia Cuevas to perhaps enjoy a certain degree of financial independence. Thus aside from reinforcing Mexican ethnicity, the process of making chorizo allowed the families to save money and simultaneously reinforce familial bonds.[17] This kind of activity also flies in the face of the notion that women were not breadwinners because they were not on a payroll.

Even for Euro-Americans living in the Southwest, and visitors, Mexican-style dishes were popular on the passenger trains of the Southwest. The demand for Mexican-style meals was met by the Santa Fe railroad's Harvey House restaurants. Harvey Houses featured *albondigas* and enchiladas in Las Vegas, New Mexico, and the *Santa Fe Magazine* published recipes.[18] While there is little evidence to suggest how many of these passengers were Hispanos or Mexicans, we may surmise that their numbers were no doubt significant given that track workers employed over six months were granted free passes. Mexican and African American customers were commonly served in separate facilities, usually behind the restaurant, or with paper sacks to be consumed outdoors.[19]

As the numbers of Mexican track workers increased and overwhelmed other ethnic groups on track work, chief cooks like Gregorio López of Santa Barbara were hired by the railroad companies to prepare meals at Mexican camps in order to further entice Mexican labor to stay. [20] Indeed Mexican and Hispano track workers walked off the job in many cases because they did not like the food they were served. [21] By 1911, the Santa Fe railroad had established a history of hiring Mexicans for track work and cultural influences flowed in both directions. For instance, J. C. Davis from Devore, California, so enjoyed Mexican food that he sang the virtues of "Tortillas con Chile." His song, written for the English-speaking employees of the Santa Fe, was published in the *Santa Fe Magazine*, along with the following description for the benefit of those unfamiliar with his Mexican delicacy; a "spoonful of the 'hot stuff' being laid upon the tortilla, the latter is rolled tightly about it. You've seen rolled jelly cake? Well, it's like that—only it *never grows cold*!"

Notwithstanding the errors in spelling, syntax and grammar, the lyrics went as follows:

¡Tortillas con Chile!

Mucho trabajo for Don Jose,
On the ferro-caril del Santa Fe;
Mucho dinaro he saves, for oh!
His corazon is in Mejico!
Now pesos are his, to pay the priest
And to buy the ring and the wedding feast;
Happy Jose! More rich than a king!
Gayly he answers the larks that sing:
"*Tortillas con chile*!"

Now, poco tiempo, Jose must go
To bring his Dolores from Mexico.
(Pobrecita! She'd die of fright
on that passajaro alone at night!)
Passes has he from the boss *padrone*—
One that "goes," for himself, alone;
One that goes—but two that "come back"—
For they need Jose on the railroad track!
What wonder the *pajaritas* call,
Loud and clear, from the chaparral:
"*Tortillas con chile*!"

A small tie casa, as white as snow,
Awaits the *chiquita* from Mexico;
And her teeth will gleam, and her eyes grow wide,
When she sees the *estufa* that stands inside;

And the *grande cajon*, whereon, each day,
She will roll *tortillas* for Don Jose—
Tortillas buenos, so fine and thin
You may plainly see, through their mellow skin,
The chile con carne, as hot as sin.
And red *frijoles*, rolled up within!
How the lark-song rings from the sagebrush gray!
A silver chime, through the golden day:
"*Tortillas con chile*!" [22]

We might safely surmise that the song writer's understanding of the importance of tortillas and chile was based on what Mexican workers told him regarding appropriate gender roles, specifically regarding tortilla preparation. It appears that he took some of their exaggerations at face value or exaggerated what was told him. Davis believed that the ability of a mexicana to pat and roll tortillas "to extreme smoothness and thinness" was "the earmark par excellence of good housekeeping."[23] Aside from the erroneous assumption that thin tortillas are actually preferable to thick ones, this song also reflects the sexist assumptions of the author (and most Mexican men at the time) who believed that women's proper place was in the home. Moreover, the song conjures up nineteenth-century images of anti-Catholicism, of Mexican "peons" oppressed by Church fees and corrupt priests. From another perspective, this song is evidence of gringo acculturation to the community of traqueros with whom they worked and with whom they occasionally socialized during off hours.

What is most important, however, is that this song reveals a very common expectation that Mexican track workers were actually very happy and expected to bring or send for their wives and family to live in the United States. This romanticized message implies that life in a tie shack with an old wood stove was an improvement for Mexicans, but it glossed over the racial subordination of Mexicans in the railroad industry. While a few Mexican immigrants may have considered a tie shack an improvement to life in Mexico, most tolerated these kinds of living conditions only until they could afford better housing.

Mutual aid societies among Hispano and Mexican immigrant workers were also important cultural institutions that were usually modeled after their counterparts in Mexico. [24] They were created and reinforced by working-class Mexicans and their families to aid one another especially during times of financial need or family crisis. Early mutualistas tended to be nationalistic, in that they reinforced a greater sense of ethnic identity among their members who felt alienated from the harsh realities of life and work in *el norte*, and who were not anxious to seek out sources of assistance from Protestant charities, hospitals, and churches. Northern New Mexico and southern Colorado probably witnessed the appearance of the first mutualistas organized primarily by traqueros. According to Sarah Deutsch, Hispanos organized seventeen in New Mexico between 1885 and 1912. La Sociedad Protección Mutua de Trabajadores Unidos (SPMDTU) formed eight of the seventeen.[25]

Other such groups included *La Cruz Azul*, *Sociedad Anahuac*, *Union Benficia Mexicana*, *Asociacion de Trabajadores Unidos*, and *Sociedad Progresista Mexicana*. However, it appears that the vast majority of Mexican track workers did not actively belong to these mutual aid societies. Nonetheless, since those involved in the leadership of the mutualistas usually assumed responsibility for organizing Mexican patriotic celebrations and festivals, it is reasonable to assume that most traqueros, like other working-class Mexicans, attended and enjoyed the cultural festivities that the mutualistas organized in the barrios.

In some areas, Mexican track workers joined and made use of the services offered by these kinds of organizations, but as discussed earlier, traqueros worked only seasonally and were highly mobile. While some mutualistas consisted of both men and women, most if not all were dominated by men, and women formed *auxiliares* (women's auxiliaries) to carry on supportive activities. In the 1890s, one of the most important mutualistas was the SPMDTU founded by Hispano traqueros in the San Luis Valley of northern New Mexico.[26]

Church groups tended not to draw the majority of traqueros into active participation for much the same reasons as mutualistas and patriotic organizations did not (seasonality, itineracy, job turnover, etc.). Nonetheless, some churches were central to the lives of Mexican track workers, but usually only in towns where there was a fairly permanent Mexican barrio. By the 1920s, many church organizations and charities began to seek out Mexican women and girls for Americanization programs in order to arrest the appeal of radical organizations, doctrines, and ideas.

Folktales and superstitions were also forms of informal cultural institutions that Hispanos, American Indian and Mexican track workers and their families carried with them when working on the railroads. While northern New Mexico and southern Colorado are rich with folktales unique to that region, many of these folktales were virtually the same as those brought by Mexican immigrants. Even so, the paucity of evidence on the work lives of Hispanos on the railroad makes it virtually impossible to know what if any folktales were prominent among Hispano track workers. Nonetheless, social scientists provide us with a glimpse of the folktales that Indian, Hispano, and Mexican track workers and their families carried with them.

For example, nighttime movement of flashing lights was long associated with witches flying and other witch-associated activity. When the Isleta Indians in New Mexico first saw the flashing lights of a locomotive off in the distance at night they immediately suspected witchcraft. According to Marc Simmons, "A light which is seen and vanishes suddenly is interpreted by both the Laguna and the Zuni as an omen of death for some loved one."[27] Elsie Parsons' investigation of the Isleta pueblo tells us that "strange lights at night are likely to be explained as witches flying from abroad to harm some townsman and that if you are anxious to harm a fellow townsman you will resort to the witches of another town."[28]

Women, more so than men, carried and transmitted folktales such as *cuentos* (stories) of "*La Llorona*" (The Weeping Woman) and "*Los Duendes*" (The Goblins). And superstitions like *el mal ojo* (the evil eye) and *brujeria* (witchcraft) were all common themes among traquero households and camps. *La Llorona*, for example, dates back to the early colonial period and is associated in most cases with a body of water. This story typically concerned a wife and mother who, after she discovered her husband with another woman, drowned her own children in a jealous rage. Murdered by her husband, she was condemned by God to spend eternity searching for her lost children along the waterways of Mexico.

Variations of the story persist to the present day, but one interesting story in Colton, California, associated *La Llorona* with the railroad. Anita Aparicio told her children that the bone-chilling cry of *La Llorona* could be heard along the Southern Pacific tracks of the South Colton barrio. [29] Adults made use of *La Llorona* to keep little ones from wandering too far away after dark, lest they be mistaken by *La Llorona* for her own children and taken away. However, the relationship of the story of *La Llorona* to the railroad tracks is not coincidental, since the majority of South Colton residents arrived as workers on the railroad during the early twentieth century. Thus some Mexicans in South Colton perhaps thought of the railroad tracks as a form of waterway that tied south Colton to old Mexico.

Besides folktales, historians Mario García and Arnoldo de Leon suggest that corridos were the most widely recognized form of oral literature that working-class Mexicans passed among themselves and their families.[30] Corridos reflected life experiences, hopes and fears, likes and dislikes, geography, world view, and environmental familiarity with the Southwest. Corridos symbolized a cultural tie that workers of Mexican background in the United States retained with Mexico. Indeed, these folk ballads reflected a bicultural experience in both Mexico and the United States, although usually with a bias toward Mexico.

Corridos in the Southwest remained Mexican in form and were sung in Spanish. However, the content often reflected the geographical and social contact with American institutions and people. In this sense, Mario García correctly points out that the immigrant corridos reflect what he terms the "immigrant experience."[31] Certainly this immigrant experience ultimately overshadowed the non-immigrant experience of Hispano railroad workers, but as is reflected in the following corrido some also reflected the Hispano experience in New Mexico during the late nineteenth century. Charles Lummis collected the following corrido in New Mexico. Note that "Franquilin" (Franklin) refers to the first railroad name for El Paso as it was known in New Mexico.

El Ferrocarril

Alli viene el ferrocarril,
Vamos á ver onde 'sta.
Ah, que gusto nos dará

¡Cuando lo veamos venir!
Coro: Llegando la emigracion
"Good morning," repetiré;
"Come in! Come in!" les diré,
Vengan oir me cancion.
De Chihuahua Franquilin

Corren los Americanos,
Ganandoles el dinero
'A todos los Mejicanos.—*Coro*.

Si fueras al campamiento
Onde vienen trabajando,
Yo asi me vivo cantando
Para ganar el sostento.—*Coro*.

Ni el sol nie el viento podrá
Hacerme retroceder;
Millas y millas correr,
Para ganar nuestro bien.—*Coro*.

La maquina va partir,
Esten toditos alerta.
Vayan sacando el dinero
Que ya vamos á partir.—*Coro*.

[translation]

The Railroad

The railroad is coming this way—
Let us go look at it near.
When we shall see it appear,
Ah, What a joy it will be.
Chorus: And when the tourists shall throng,
"Good morning!" I will repeat;
"Come in! Come in!" I'll entreat,
"Come ye and list to my song!

Up from the town on the line
come running the Americanos,
Earning us everyone money—
Money for all us paisanos.—*Chorus*.

Were you at the camp where they're giving
Work, and the laborers bringing—
That's just the way I live singing

Only to earn me a living.—*Chorus.*

Neither sun, neither wind shall nor could
Make me turn back till I've done;
Mile after mile I will run
That I may win us some good.—*Chorus.*

The engine is going to start.
Lively! Be all of you ready!
Come, pull your money out—steady,
For now we are going to start!—*Chorus.*[32]

It should be kept in mind that the term *campamiento* (encampment) refers to a temporary camp usually located at a work site, and should not be mistaken for a barrio, although some encampments may have evolved into barrios later.

Another corrido by the same name illustrates the immigrant experience and lays out the fears of leaving Mexico and never returning.

El Ferrocarril

La máquina pasajera
no puede hacer cosa buena
porque "oscurece" en su casa
y amanece en tierra ajena

¡Ay! ¡que dolor!
Tendrían los mexicanos
al ver el ferrocarril
que traen los americanos.
La máquina chiquitita
es la que ha quedado aquí

y la quieren llegar
has San Luis Potosí
Oigan y oigan
el ferrocarril bramar.
el que lleva a los hombres
y nunca los vuelve a traer.

[translation]

The Railroad

The fleeting engine
Can't do anything good
Because at dusk it is at home
And at dawn in a strange country.
Oh! What sadness!
The Mexicans will have to see
The railroad train
That the Americans bring.
The very littlest engine
Is the one that has been left here,
And they expect it to go
As far as San Luis Potosi.
Listen, Listen,
To the train puffing;
The train which carries men away
And never brings them back again.[33]

The fears of not seeing one's family again were peculiar to first-time working-class Mexican immigrants and parallels other classic immigrant experiences, whether on the railroads or other employment sectors. This may suggest why many if not the vast majority of working-class Mexican immigrants preferred railroad work at first, since the railroads offered free return passage after six months of service. The following corrido also exemplifies the fears associated with contracted labor agreements in that the *enganchadores* (contractors) did not inform traqueros of their destination, and refused responses when workers inquired. This practice may have been to prevent workers from jumping off the train if they knew where they were going.

Los Enganchados

El 28 de Febrero,
Aquel día tan señalado,
Cuando salimos de El Paso
Nos sacaron reenganchados.

Cuando salimos de El Paso
A las dos de la manana,
Le preguntó al reenganchista,
si vamos para Louisiana.

Llegamos a la Laguna
Sin esperanza ninguna
Le pregunté al reenganchista
Se vamos para "Oclajuma."

Por esas lineas del Kiri [Katy]
Pasa un tren muy volador
Corre cien milla por hora
Y no le dan todo el vapor.

Y el que no lo quiera creer
Nos más que venga a montar
No más que monte de noche
Lo verá donde va a dar.
Llegamos el día primero
Y al segundo a trabajar
Con los picos en las manos
Nos pusimos a trampar.

Unos descargaban rieles
Otros descargaban "tallas"
Y otros de los compañeros
Echaban de mil malallas.

Los que sabían el trabajo
Iban recorriendo el "llaqui"
Martilleros y paleros
Echándole tierra al traque.
Ocho "varas" alineadas
Nos seguíamos disgustados
los gritos y las señas
Nos quedábamos paraos.

Decía D. Jose Maria
Con su boquita de infierno

Mas valiera estar en Kansas

Que nos mantenga el gobierno.
Decía Jesus el Coyote
Como queriendo llorar
Valía más estar en Juarez
Aunque sea sin trabajar.
Estos versos son compuestos
Por un pobre mexicano
Pa ponerlos al corriente
Del sistema americana.

[translation]

The Hooked Ones

One the 28th day of February,
that important day
When we left El Paso,
They took us out as contract labor.
When we left El Paso
At two in the morning,
I asked the boss contractor
If we were going to Louisiana.
We arrived at Laguna
Without any hope.
I asked the boss
If we were going to Oklahoma.
Along the line of the Katy
There goes a very fast train.
It runs a hundred miles an hour
And then they don't give it all the steam.
And he who doesn't want to believe it,
Just let him get on board.
Just let him get on board at night;
He will see where he gets to.
We arrived on the first day
And on the second began to work.
With our picks in our hands
We set out tamping.

Some unloaded rails
And others unloaded ties,
And others of my companions
Threw out thousands of curses.
Those who knew the work
Went repairing the jack
With sledge hammers and shovels,
Throwing earth up the track.
Eight crowbars lined up,
We followed disgusted;
to shouts and signs
We remained indifferent.
Said Don José Maria
With his hell's mouth,
"It would be better to be in Kansas
Where the government would maintain us."
Said Jesus, "El Coyote,"
As if he wanted to weep,
"It would be better to be in Juarez
Even if we were without work."
These verses were composed
By a poor Mexican
To spread the word about
The American system.[34]

Aside from the fears associated with the enganche, this corrido clearly illustrates the various operations that track workers performed such as unloading rails and ties; repairing machinery; tamping the earth around the ties; and the collective process required to move rails with crowbars in cadence to the timed calls of the foreman.

The theme of the enganche is a common one since many working-class Mexican immigrants were first "hooked" by enganches when they initially arrived in the border cities. Paul S. Taylor collected the following verse during the 1920s. The title is similar to the previous corrido, but also reveals the cultural transformation from peasant status to industrialized worker. One oldtimer recalls that the standard attire from rural agricultural areas was the *calzon blanco* (white cotton trousers and shirt):

El Enganchado

Desde Morelia vi enganchado.
Garnarlos dolares fue mi ilusión.
Compre zapatos, compre sombrero.
Y hasta me puse pantalon.

[translation]

The Hooked One

I came from Morelia to work under contract.
To make lots of money was my dream.
I bought some shoes, and I bought a hat.
And now I'm even wearing trousers.[35]

Soon after arriving in the United States, the *calzon blanco* was discarded for American denim overalls and other work clothes. [36] The following corrido is one of the few tunes that imitates the "tickety-tackety, tickety-tackety" sound of a car rapidly traveling over the rails, according to Américo Paredes:

La Pensilvania

El dia veíntiocho de abril,
a las seis de la manana,
salimos en un enganche
para el estado de Pensilvania.
Mi chinita me decia:
—Yo me voy en esa agencia,
para lavarle su ropa,
para darle su asistencia.—
Y el enganchista nos dice:
—No se lleven la familia,
para no pasar trabajos
en el estado de West Virginia.—
—Pa' que sepas que te quiero,
me dejas en Foro West;
cuando ya estes trabajando
me escribes de donde estes.

—cuando ya estes trabajando
me escribes, no seas ingrato;
de contestacion te mando,
de recuerdo, mi retrato.—
Al llegar a West Kentockle
cambiamos locomotora,
de alli salimos corriendo
ochenta millas por hora.
Adios Foro West y Dalas,
pueblos de mucha importancia;
ya me voy pa' Pensilvania
por no andar en la vagancia.
Adios estado de Texas,
con toda su plantacion;
ya me voy pa' Pensilvania
por no pizcar algodon
cuando llegamos alla,
que del tren ya nos bajamos,
nos dicen las italianas:
¡Di' onde vienen, mexicanos?—
Contestan los mexicanos,
los que ya sabian ingles:
—venimos en un enganche
de la ciudad de For West [*sic*].—
Estos versos son compuestos
cuando y venia en camino,
poesias de un mexicano
de nombre Constantino.
Ya con esta me despido,
con mi sombrero en las manos,
y mis fieles companeros
son trescientos mexicanos.

[translation]

The Pennsylvania

On the twenty-eighth day of April,
at six o'clock in the morning,
we left on a labor contract

for the state of Pennsylvania.

My beloved said to me,
"I want to go on that job,
so I can wash your clothes,
so I can cook your meals."

But the labor contractor told us,
"Don't take your families along,
so you won't have any trouble
in the state of West Virginia."

"Just to show you that I love you
leave me, then, in Fort Worth;
when you are settled and working,
write me, wherever you are."

"When you are settled and working,
please write me, don't be mean;
when I answer I will send you my picture
to remember me by."

When we got to West Kentucky,
we switched locomotives;
we went out of there going at
eighty miles per hour.

Farewell, Fort Worth and Dallas,
you very important towns,
I'm going to Pennsylvania
to avoid a vagrant's life.

Farewell, you state of Texas,
with all of your planted fields,
I'm going to Pennsylvania
so I won't have to pick cotton.

When we arrived there,
once we had got off the train,
the Italian girls asked us,

"Mexicans, where are you from?"

And the Mexicans answered,
those who already knew English,
"We have come on a labor contract
from the city of Fort Worth."

These stanzas were composed
while I was still on the way;
they're the verses of a Mexican
by the name of Constantino.

Now with this I say farewell,
with my hat in my hands,
and my faithful companions
are three hundred Mexicans.[37]

As suggested by the corrido, families were not encouraged to accompany seasonal railroad employees due to any number of reasons. In this case, it appears that some people in West Virginia did not mind if Mexicans were brought in to work on the tracks as long as they left and did not settle there with their families.

Unlike cuentos and dichos, corridos tended to be culturally transmitted by men instead of women. While men typically sang these songs with and among their families, they also sang these corridos to themselves at home, in the work camps, and especially in *cantinas* (saloons or bars). Indeed, singing in bars was usually made more boisterous with alcoholic consumption. And while alcohol probably made for loud and more boisterous singing, the general tradition of singing in the company of men was widespread—among both Hispano railroad workers and Mexican immigrant track and railroad workers—and was done freely with or without alcohol.

Singing corridos and playing musical instruments (usually guitars or violins) whether at home or outside the home, was also a common feature of large family gatherings, parties, weddings, anniversaries, and fiestas patrias.[38] According to Manuel Gamio, corridos may be best defined as "songs which tell a story. The corridos are of all the songs collected, nearest to the human interest story of the popular newspaper. Like the human-interest story, they express the interests and attitudes of the people."[39]

Railroad officials, social workers, and government officials generally categorized Mexican track workers as "casual labor" or "common labor." These tens of thousands of working-class immigrants were in turn targeted by social workers and xenophobic public officials for Americanization programs beginning with World War I.

The "Mexican Problem" or "Mexican Question," as politicians and newspapers termed it, was rooted in the idea that Mexicans posed a potential cultural threat to American society because they were perceived as "unassimilable" or uninterested in adopting American culture. The large numbers of track workers of Mexican background also alarmed many observers due to the real or imagined health conditions that surrounded Mexican track worker communities. Since immigration has historically been associated with foreign ideas and radical political doctrines, it was particularly important for the corporate interests to Americanize the Mexican community in order to forestall groups like the Industrial Workers of the World and the Partido Liberal Mexicano from gaining a toehold in the Mexican community.[40]

Track work was among the most difficult and least desirable, but it was probably also the easiest work for working-class Mexicans to find. This employment sector tended to attract first time immigrants, usually "unskilled" single young men.[41] For this reason railroad officials were also interested in Americanizing Mexican casual laborers in order to increase their company loyalty and ultimately embrace the assimilationist ethos. As George J. Sánchez has recently shown, Mexican women were singled out by Americanization agents who believed that the best way to Americanize Mexican families, and especially children, was to reach the women first. They believed that if women could be won over to Americanization than they would in turn influence the home. After 1920, Americanization efforts moved away from the homes and labor camps to the schools and community centers. In 1929 Pearl Idelia Ellis, an Americanization expert in Southern California, wrote in *Americanization Through Homemaking* that the "influence of the home" was important in shaping the worker who, according to Ellis, "is more dependable and less revolutionary in his tendencies." Women, or the "homekeeper creates the atmosphere, whether it be one of harmony and cooperation or of dissatisfaction and revolt."[42] Another observer of Americanization efforts promoted the company-owned labor camps for Mexican employees. He wrote:

> In order to make the labor camp pay—to make it produce the desired workers—it is necessary to create an atmosphere that will attract and hold such workers. The Mexican is reached most easily through his home and friends. If his family is well and happy, if his house is neat and clean, and his meals good—and above all if his yard is full of vegetables and flowers which he has planted—he will be pretty certain to think twice before uprooting himself. [43]

Thus with a full vegetable garden, well prepared meals, and a clean home, Mexican track workers were expected to behave and perform efficiently and presumably with no interest in unionization activity. And women, of course, constituted the unpaid work-force behind Americanization that social workers and the captains of industry hoped would be the catalyst behind the program.

According to Sánchez: "in a study conducted in 1923, 55 percent of the Mexican immigrants surveyed considered it their duty to remain politically loyal to Mexico, while almost all of the rest refused to answer the question."[44] Americanizers dismissed the possibility of transforming "naturalized foreigners." Agents of Americanization argued that this Immigrant Generation: "are never 100% Americans, but the second generation may be. 'Go after the women' and you may save the second generation for America." [45] By implication then, Americanizers thought that working-class Mexican women were understood to be more open to Americanization than men were, and less inclined to retain a Mexican ethnic identity and cultural values.

An underlying theme of Americanization classes reinforced hierarchical gender roles. In 1912, some traqueros in East Chicago attended Americanization classes in evening schools. With the European immigration restriction after 1917, the number of Mexican immigrants in East Chicago and elsewhere in the Midwest grew. By 1926, there were close to 500 Mexicans enrolled in twenty-one classes in Hammond Lake, Illinois. Men were given magazines and newspapers to read in order to discuss current issues and women were placed in "Home Helps" classes where they were taught English by discussing "home problems." [46] In 1919, *La Mision Mexicana*, a Protestant evangelical mission in Sheffield, Kansas, offered classes in English and sewing for mexicanas.[47] Thus, intellectual knowledge was stressed for men and domestic knowledge was stressed for women. This gendered arrangement parallels the idea of separate spheres discussed by Mary Ryan in her discussion of women in Oneida County, New York. [48]

In 1915, California passed the Home Teacher Act, which allowed school districts to employ teachers to familiarize pupils and parents with American culture. Americanizers organized classes to teach English, American cooking, housekeeping, sewing, dancing, singing, sanitation, and games, among other things, to immigrants. The basic underlying idea behind Americanization classes was, according to one Americanization expert, to "acquaint everyone in our country... with the truths and duties at the base of American society" and to "introduce the foreign-born to America and explains America to the foreign-born."[49]

Perhaps one of the most telling definitions of Americanization is one given by the Secretary of the Interior: "we want the people who live here to feel a responsibility for the conditions that exist here and that their complaint is against themselves, not against another class [emphasis added] in the community. [50] Thus class consciousness, especially of the kind promoted by groups such as the Industrial Workers of the World and other radical organizations, was simply un-American. Americanizers taught Mexican track workers, like all workers, that they individually were to blame for social inequality, not the burgeoning system of industrial capitalism.

To further Americanization specifically among railroad workers, the Railroad YMCA represented a special arm of the YMCA with its own program and funds donated directly by the railroads and the workers. Its objectives were identical to that of Americanizers: to teach immigrant railroad workers how to speak English through reading and religious instruction, especially in the Bible. Moreover, according to Servando I. Esquivel, a Mexican American leader of the YMCA:

> Its [the YMCA] efforts tend to increase the efficiency of the workmen; to avoid economic disturbances; and, above all, to promote a better understanding between Americans and Mexicans, through the interpretation of that which is good and worthy in life of both peoples.[51]

Underlying Esquivel's statement are subtle but important reflections of the major issues of the Progressive Era, these included: "efficiency" and scientific management; "economic disturbances" and the fear of radical labor actions (strikes, boycotts, sabotage, etc.); "understanding between Americans and Mexicans" referring to the pressing racial ethnic inequality related to Mexican immigration.

In September 1920, a young Mexican American by the name of S. L. Hernández was hired to do Americanization work among the Mexican track workers of the Santa Fe railroad in Albuquerque. Esquivel, on the other hand, was an educated spokesperson for the Asociacíon Cristiana de Jovenes (YMCA) in El Paso and was in charge of Americanization work among the thousands of Mexican track workers in El Paso, Tucson, Metcalf, and Miami, Arizona. Esquivel reasoned:

> helping to increase the wealth of this country, they [Mexican track workers] are entitled to something more than wages. We mean by this better housing conditions, educational opportunities, and health and recreational advantages. For, after all, no real prosperity can be accomplished at the expense of manhood. [52]

By implication, we might conclude that Esquivel believed that it was perfectly acceptable to accomplish prosperity at the expense of womanhood, since from this perspective women were regarded as non-workers. Nonetheless, despite Esquivel's efforts with the YMCA, it appears that the YMCA had only nominal impact on the lives of Mexican track workers. Still while the available evidence does not reveal the numbers of Mexican workers that the YMCA was able to assist, it is probably reasonable to assume that some Mexican traqueros took advantage of at least some of the basic services at one time or another.

On the evening of February 2, 1911, nearly 200 Mexican (along with a few Japanese and Greek) track workers attended an English lesson by the Railroad Young Men's Christian Association (YMCA) in Argentine, Kansas. The YMCA hired Dr. Peter Roberts from New York to teach the group English by outlining the steps for "getting up in the morning." [53]

In 1921, approximately seventy men in Emporia, Kansas, attended an English class held at Central School. However, most stopped attending when the Protestant teacher made anti-Catholic statements. The Santa Fe railroad also provided space for English classes for its Mexican workers in Emporia, but these classes were also short-lived.

Local Americanizers did not offer an explanation for the apparent unpopularity of the classes in Emporia. [54]

In a letter to A. J. Fink, regarding the Americanization of its Mexican workers of the Rock Island Railroad, an O. A. Morse boasted that:

> Both the employe [*sic*] himself as well as his wife and children are included in this educational work and our efforts are beginning to bear their fruit. This is indicated by the neat condition of their home and the greater interest that is being taken in their work. [55]

Morse and other Rock Island Railroad officials erroneously believed that Mexican track worker families were incapable of keeping neat homes and working conscientiously without Americanization.

Of course, Americanization programs were not limited to railroad-workers. Indeed, they were present wherever Mexican workers were concentrated. For example, women and girls in some agricultural labor camps in California were taught to sing songs about the work they performed. Upon graduation from class, according to one Industrial Relations official, students were made to demonstrate what they learned. A representative of the Industrial Relations Bureau said:

> On a table before the successful students there were miniature beds, wash pans, brooms, dolls and other household paraphernalia. Each student said her little piece in frightened English, demonstrating it by making the bed, dressing the doll or sweeping the floor. They sang songs in English about opening the windows, washing the baby and learning English. [56]

The uniformity of Americanization programs among Mexican workers across one industry to the other was striking, although not exactly identical. Programs ranged from English classes to elaborately organized classes with ongoing activities through settlement houses, the YMCA, the YWCA, churches, and schools. However, the ultimate objective was the same: to teach American culture to Mexicans in order to win their loyalty to the United states and its institutions as a way of controlling labor relations.

Cultural relationships between Mexican railroad workers and their families both at home and on the job were complex and fluid. Nonetheless, Mexican railroad worker culture clearly evolved as a distinct part of an American working-class culture which was inextricably tied to their daily lives as both workers and residents on or near the railroad yards. While the experiences and behavior of Mexican railroad workers and their families were not uniform, there were important cultural expressions that speak to the adaptability and resiliency of Mexican men and women. From the 1880s to the Depression, Hispano and Mexican immigrants adjusted themselves to the new conditions of industrial life, and their contact with Euro-American institutions—especially schools—slowly transformed Hispanos and Mexicans into what we might call traquero culture. Mexican railroad worker culture was similar to that of Mexican workers in other industries, but differed in that the experience of actual track work and shop work shaped their particular relationships. Although Hispano and Mexican immigrants did not control the formal institutions that would have allowed them to retain control of their traditional lands and communities, they nevertheless negotiated and adjusted to rapidly changing conditions imposed by the railroad. At the same time Mexican railroad workers underwent cultural change, cultural continuity was also manifested in the everyday family, community, religious, and recreational activities.

conclusion

THROUGHOUT THE LATE NINETEENTH AND EARLY TWENTIETH centuries, Mexican workers have been inextricably linked to the development of the Southwest. The mining, ranching, farming, timber, and manufacturing industries of this region all profited by the abundance of inexpensive labor from Mexico. Railroads served as the conduit for the recruitment and transportation of Mexican workers to these industries not only in the Southwest but in virtually every section of the country as well. For Mexican workers, more than any other immigrant workers, the railroad linking Mexico with the United States represented the first and the largest industrial employment on either side of the border.

As America's "first big business," railroads themselves became huge consumers of steel, coal, lumber, timber, and other raw materials, thereby stimulating the economy and creating new jobs. Indeed, the railroad was also Mexico's "first big business."[1] The connection between American railroads and Mexican development underlay the social displacement and migration of Mexican workers within Mexico and into the United States. In 1880, with the permission of the Mexican government, the Atchison, Topeka & Santa Fe Railroad organized the *Ferrocarril Central de Mexico* and the *Ferrocarril de Sonora*. And, the Denver & Rio Grande Railroad organized the railroad which eventually became the *Ferrocarril Nacional de Mexico*. By the turn of the century, these two American railroads controlled eighty percent of all investments in Mexican railroads.[2] Although American railroads organized their first crews in the United States, they eventually hired Mexicans for track work and other unskilled and semi-skilled jobs in Mexico. Soon after, American labor recruiters promised better pay and lured many of these workers north of the border to take jobs on American railroads, often for half the wages paid to white native-born workers.

Mexican labor on America's railroads is only part of the larger process of the great waves of immigrant labor from Europe and Asia. Like Chinese, Irish, Italian, Greek, and eastern European immigrants, Mexican immigrants helped to build the transportation infrastructure for America's growth. But, unlike European immigrants whose flow from Europe was curtailed by the 1917 Immigration Act, the federal government exempted Mexican immigrants from these restrictions. Thus Mexican immigration continued into the 1920s even though the major trunk lines were completed and the technological advances in track maintenance required fewer track workers. Additionally, transportation by automobile and trucking began to cut into shipments by railroad thus reducing tonnage, which led to less track maintenance and fewer traqueros. Nonetheless, railroads continued to complain about the seasonal labor shortage for maintenance-of-way work.[3]

Mexican employment on the railroads began prior to the first transcontinental linkage between the United States and Mexico in 1881. In 1871, native-born Mexicans (Hispanos) of southern Colorado and northern New Mexico found employment on the grade and the railhead of the Atchison, Topeka & Santa Fe Railroad mostly as track hands. Then, as the Santa Fe Railroad and the Southern Pacific Railroad linked Mexico with the United States in the spring of 1881, Mexican immigrant workers slowly ventured into the southwestern United States and beyond. The 1882 Chinese Exclusion Act, which coincided with the Southern Pacific and Santa Fe Railway link to Mexico, curtailed Chinese immigration.

Immediately, a few Mexican workers ventured north into the Central Plains, the West, the Pacific Northeast, Midwest, and even parts of the South. However, the great cyclical flow of migrant Tejano laborers north and Mexican immigrants into the United States accelerated after the turn of the century. Nonetheless, Mexican origin workers sought gainful employment in virtually every state in the union. The demand for Mexican labor continued throughout the 1920s, dropping toward the end of the decade as a result of President Hoover's anti-Mexican repatriation campaign. With the onset of World War II, the United States again looked to Mexico to help supply its demand for labor. The Railroad Bracero program was one branch of the wartime bi-national agreement to supply American industries with "helping arms" from Mexico.[4]

Among the many advantages to hiring Mexican labor, the railroads appreciated Mexican workers most for their willingness to accept wages and working conditions that white native-born workers would not. However, for mexicanos, their high geographical mobility was the most striking feature of their lives in the United States. Not only did they move with great frequency from one job site to another, for the most part they also moved from one occupation to another, and then returned to Mexico only to turn around for another work stint in the United States. The seasonal itinerary for one traquero from Arandas, Jalisco illustrates this fluidity:

> [In] 1910, [he] worked for the Southern Pacific in New Mexico, and returned to Arandas in 1911; went to the United States in 1913, worked for the Santa Fé in Kansas, and returned to Arandas in 1914; went to the United States in 1916, worked on a pipe line being laid from Kansas City to Chicago, and returned to Arandas the same year; went to the United States in 1918, worked on the track and in cotton picking, and returned to Arandas, in 1920; went to the United States in 1922, worked in the Santa Fé yard in Kansas City, and returned to Arandas in 1924; went to the United States in 1925, worked for the Southern Pacific at Phoenix, Arizona, and returned to Arandas, in 1926; went to the United States, in 1928, worked for the Rock Island at Denver, and returned to Arandas in 1929.[5]

Indeed, for particularly ambitious Mexican workers, an exhausting six-month work stint on the tracks convinced them to find other better paying and less strenuous employment. Many of these workers found employment in Detroit and the industrial Midwest in the auto industry, foundries, and steel manufacturing. Some found employment in the railroad shops where their chances of living among friends and family were greater.

Once family, friends, and neighbors established themselves in various parts of the United States, they sent for relatives, *compadres*, and friends and found them places to live and work. This chain migration and chain occupational network allowed recent arrivals to obtain year-round section work, as opposed to seasonal employment.

By the mid-twenties, many immigrants by-passed the back-breaking track work altogether. Additionally, chain migration gave rise to clear patterns of migration and settlement between specific towns in the United States and towns in sending states of central Mexico.

Track work itself took a terrible toll on the health of traqueros. Injury and death due to accidents and negligence were commonplace on the railroads. Moreover, the proximity of traquero family housing in boxcar communities also meant that women and children not employed by the railroads were also occasionally maimed or killed near the railroad tracks. Of course, these kinds of accidents were not limited to Mexicans; however, since Mexicans tended to be housed at the railroad yards or near the round-houses, the frequency seemed greater.

Traqueros also lived in a variety of housing situations, from housecourts in Los Angeles, boarding houses in Chicago, rolling tent villages in Texas, and shanties and boxcars in the Midwest, Central Plains and Southwest. The introduction of scientific management reinforced family reunification and patriarchy. The formation of Mexican ethnic enclaves along the railroad lines, especially in the Midwest and Central Plains, created a form of "temporary nationalism" in traquero communities. The common experiences of immigration, track work, racial ethnic discrimination, and segregation in housing reinforced Mexican cultural and linguistic boundaries. At the same time, this cultural boundary maintenance allowed for immigrant Mexicans to acculturate to American customs and work norms in a familiar environment.

Paul S. Taylor of the University of California, Berkeley, provided an important insight about Mexican labor in the United States during the first few decades of the twentieth century, when he wrote: "Employment by railroads on maintenance of way, work in round-houses, etc., was by far the most common experience of the emigrants."[6] On new roads traqueros prepared the grade, laid ties and rails, dug ditches, tunneled through mountains, and built trestles, among other things. Once the railroad was completed, they replaced old ties and rails, repaired roadbeds, weeded, tightened bolts, and repaired switches and frogs, to name just a few of their many responsibilities. Yet, railroads insisted that this work be classified as "unskilled," and thus track work remained the lowest paid in the railroad employment hierarchy.

Occasionally, mexicanas found employment on the tracks along with their husbands and male family members. However, the majority of Mexican women worked at home or in domestic-service-related industries, as cannery workers, cooks, and housekeepers. Others participated in the informal economy and prepared lunches for neighboring single men, took in boarders, or sold Mexican food on street corners. And, of course, some traqueros abandoned their wives and children in Mexico and caused them greater hardship. However, many more women traveled with their husbands and menfolk and worked alongside them either on the tracks, or at home, and helped to rebuild Mexican families and communities in the United States.

The present study has built upon the work of Chicano and labor historians, who have contributed to our knowledge of America's working-class and to the largest and oldest ethnic group in the southwestern part of the United States. However, much of this recent scholarship has tended to focus on the regional or community studies. This work has attempted to build upon the foundation of this fine tradition of scholarship and hopefully to transcend it. *La traquera* (track work) was perhaps the most common occupational experience among all immigrant Mexican workers between 1871 and 1930. By examining traqueros, probably the lowest and most fluid sector of the railroad occupational hierarchy, this study strives to reconstruct the daily lives of this invisible labor force across regions and communities. While this study has not examined each and every region of the country, the attempt herein is to explore this national experience for Mexicans workers in the United States.

While the first traqueros were the descendants of the nineteenth-century American conquest of Mexican territory, these *Hispano* traqueros were soon greatly outnumbered and pushed out of track labor positions by immigrant traqueros. Mexican track workers were by no means homogeneous; however, as a group they represented a wide variety of Mexican working-class culture in the United States. Indeed, Mexican-origin workers remain the largest and oldest Latino group in the United States. This study helps us to see that Chicanos perhaps more so than any other Latino subgroup are a national minority group, and not merely a regional minority—one which has heretofore been largely ignored in terms of their contribution to the building of America's railroads.

Endnotes

Notes to Foreword

I would like to thank the following colleagues who so generously shared their memories of Jeffrey Garcilazo's life and work: María Raquél Casas, Mario García, Gabriel Gutiérrez, Heidi Tinsman, Steven Topik, and Rodolfo Torres. I am also grateful to Valerie Matsumoto for her thoughtful editorial skills.

1. Gregory Rodríguez, "Why Arizona banned ethnic studies," *Los Angeles Times* (February 20, 2012); Jeff Biggers, "Tucson says banished books may return to classroom," *Salon* (January 18, 2012); Jeff Biggers, "Who's Afraid of 'The Tempest'?," *Salon* (January 13, 2012); *In the Matter of the Hearing of an Appeal by the Tucson United School District*, Case Number 11F-002ADE, Office of the Administrative Hearings (December, 27, 2011) Available at http://azstarnet.com/news/local/education/precollegiate/pdf-tusd-ethnic-studies-ruling/pdf_adf0b17a-52a0-11e1-8ac0-0019bb2963f4.html
2. See for example, Lilia Fernández, *Brown in the Windy City: Mexicans and Puerto Ricans in Postwar Chicago* (Chicago: University of Chicago Press, forthcoming); Gabriela Arredondo, *Mexican Chicago: Race, Identity, and Nation, 1916–39* (Urbana: University of Illinois Press, 2008); Dionicio Nodín Valdés, *Barrios Norteños: St. Paul and Midwestern Mexican Communities in the Twentieth Century* (Austin: University of Texas Press, 2000); and Zaragosa Vargas, *Proletarians of the North: A History of Mexican Industrial Workers in Detroit and the Midwest, 1917–1933* (Berkeley: University of California Press, 1999).

3. Interview with Ray Buriel, December 21, 1994, conducted by the author as quoted in *From Out of the Shadows: Mexican Women in Twentieth-Century America* (New York: Oxford University Press, 1998): 22.

Notes to Introduction

1. Clippings, Kansas State Historical Society, Topeka (hereafter KSHS). I am very grateful to Connie Menninger at the KSHS for her assistance and cooperation with this project.
2. A word about terminology is needed. The term "Mexican" is preferred in the title throughout this study as opposed to others for the following reasons: "Mexican" historically encompasses Hispanicized Indians and Hispanos, whereas the others do not; all three tend to be lumped together as such in the documents; and most importantly, Mexican immigrants (or migrants, as some have argued) far outnumbered Indians and Hispanos on the railroad, especially in track work. Nonetheless, each group will be identified as Indian, Hispano, or Mexican when necessary in order to distinguish between them. The term "mexicano" (Spanish pronunciation of Mexican) is herein used interchangeably with Mexican. The term "Indian," or "American Indian" is herein used to refer to those who identified themselves as Indian and whom others identified as Indian and not Mexican. The term "Hispano" is used in this study to designate those "Mexicans" from the regions of southern Colorado and northern New Mexico who preferred to identify themselves as such. The term "Hispano" is also used interchangeably with "Mexican American" (U.S.-born) as opposed to traqueros born in Mexico. The terms "Anglo," "white" and "Euro-American" are herein used interchangeably. And although these terms clumsily lump together people from Europe, specific European ethnics are designated when the data allows it.
3. Chicano historians who have included discussion of Mexican track work are: Albert Camarillo, *Chicanos in a Changing Society: From Mexican Pueblos to American Barrios in Santa Barbara and Southern California, 1848–1930* (Cambridge: Harvard University Press, 1979); Mario T. García, *Desert Immigrants: The Mexicans of El Paso, 1880–1920* (New Haven:

Yale University Press, 1981); Ricardo Romo, *East Los Angeles: History of a Barrio* (Austin: University of Texas Press, 1983); Mario Barrera, *Race and Class in the Southwest: A Theory of Racial Inequality* (Notre Dame: University of Notre Dame Press, 1979); other works have added more to our understanding of the experiences of Mexican track workers, for example: Michael M. Smith, *The Mexicans in Oklahoma* (Norman: University of Oklahoma Press, 1980); Thomas E. Sheridan, *Los Tucsonenses: The Mexican Community in Tucson, 1854–1941* (Tucson: University of Arizona Press, 1992); George J. Sánchez, *Becoming Mexican American: Ethnicity, Culture, and Identity in Chicano Los Angeles, 1900–1945* (New York: Oxford University Press, 1993); Maria Herrera-Sobek, *Northward Bound: The Mexican Immigrant Experience in Ballad and Song* (Bloomington: Indiana University Press, 1993); Zaragosa Vargas, *Proletarians of the North: A History of Mexican Industrial Workers in Detroit and the Midwest, 1917–1933* (Berkeley: University of California Press, 1993); and especially the monograph by Emilio Zamora, *The World of the Mexican Worker in Texas* (College Station: Texas A&M University Press, 1993); see also Tomás Almaguer, *Racial Faultlines: The Historical Origins of White Supremacy in California* (Berkeley: University of California Press, 1994).

4. Mario T. García, *Mexican Americans: Leadership, Ideology, and Identity, 1930–1960* (New Haven: Yale University Press, 1989): 1–22, 293–302.
5. This analysis is borrowed from David Montgomery, *Fall of the House of Labor: The Workplace, the State, and American Labor Activism, 1865–1925* (Cambridge: Cambridge University Press, 1987): 55–111.
6. L. L. Waters, *Steel Rails to Santa Fe* (Lawrence: University of Kansas Press, 1950): 287.
7. James Marshall, *Santa Fe: The Railroad that Built an Empire* (New York: Random House, 1945): 180–1, 189.
8. See Paul S. Taylor's seven-volume study of *Mexican Labor in the United States* (Berkeley: University of California Press, 1929–1934); and Victor S. Clark's "Mexican Labor in the United States," *Bulletin of the Department of Labor* 78 (September 1908): 469–74. See also Luis Leobardo Arroyo, "The State of Chicano Labor History 1970–1980," in *Chicanos and the Social Sciences: A Decade of Research and Development (1970–1980)*,

ed. Isidro D. Ortiz (Santa Barbara: Center for Chicano Studies, University of California, 1983): 1–8.

9. Walter Licht, *Working for the Railroad: The Organization of Work in the Nineteenth Century* (Princeton, New Jersey: Princeton University Press, 1983); and Montgomery, *The Fall of the House of Labor*, 58–111.

10. Montgomery, *Fall of the House of Labor*, 58.

11. Judith Laird, "Argentine, Kansas: the Evolution of a Mexican-American Community, 1905–1940" (Ph.D. dissertation, University of Kansas, 1975); other less ambitious studies are Cynthia Mines, "Riding the Rails to Kansas: The Mexican Immigrants" (M.A. thesis, McPherson College, 1980); Robert Oppenheimer, "Acculturation or Assimilation: Mexican Immigrants in Kansas, 1900 to World War III," *Western Historical Quarterly* (October 1985): 429–48; Charles Wollenberg, "Working on El Traque: The Pacific Electric Strike of 1903," *California Historical Quarterly* 42 (August 1973): 358–69.

12. See Michael M. Smith, *The Mexicans in Oklahoma* (1980); "Beyond the Borderlands: Mexican Labor in the Central Plains, 1900–1930," *Great Plains Quarterly* (Fall 1981): 239–51; "The Mexican Immigrant Press Beyond the Borderlands," *Great Plains Quarterly* (Spring 1990): 71–85; "Mexicans in Kansas City: The First Generation, 1900–1920," *Perspectives on Mexican American Studies* (1989): 29–57.

13. Alfred D. Chandler, *The Railroads: The Nation's First Big Business* (New York: Harcourt, Brace and World, 1965).

14. By "subaltern" I refer to the subordinate and seemingly passive forms of resistance. See chapters 3 and 4.

Notes to Chapter 1

1. See the published work: Deena González, *Refusing the Favor: The Spanish-Mexican Women of Santa Fe, 1820–1880* (New York: Oxford University Press, 1999). Quoting from page 107 of the manuscript with permission from Deena González. I am also grateful to Professor González for her assistance and feedback through the Tomás Rivera Center.

2. García, *Desert Immigrants*, 15–16; see also, D. W. Meinig, *Southwest: Three Peoples in Geographical Change 1600–1970* (New

York: Oxford University Press, 1971), 38–39; for Cimarron Cutoff see Alvin R. Sunseri, *Seeds of Discord: New Mexico in the Aftermath of the American Conquest, 1846–1861* (Chicago: Nelson-Hall, 1979): 70; Thomas D. Hall, *Social Change in the Southwest, 1350–1880* (Lawrence: University of Kansas Press, 1989): 149.

3. García, *Desert Immigrants*, 15–16.
4. Owen White, *Out of the Desert: The Historical Romance of El Paso* (El Paso: McMath, 1923): 124.
5. Thomas A. Rickard, *A History of American Mining* (New York: Johnson Reprints, 1932): 27.
6. See Richard Griswold Del Castillo, *The Treaty of Guadalupe Hidalgo: A Legacy of Conflict* (Norman: Oklahoma University Press, 1990), passim; White, *Out of the Desert*, 74, 177–78.
7. *Las Vegas Daily Optic* (hereafter *Optic*), 9 April 1881.
8. Roscoe Conkling and William D. Shipman, *The Central Pacific Railroad Company in Equitable Account with the United States, Growing out of the Issue of Subsidy Bonds in Aid of Construction, A Review of Testimony and Exhibits Presented Before the Pacific Railway Commission, Appointed According to the Act of Congress, Approved March 3d, 1887* (New York: Henry Bessey, 1887): 59, hereafter *Pacific Railway Commission*. Also cited in Yen Tzu-Kuei, "Chinese Workers and the First Transcontinental Railroad of the United States" (Ph.D. dissertation, St. John's University, 1977): 117.
9. Ibid., 60.
10. Blanche C. Grant, *When Old Trails Were New: The Story of Taos* (Glorieta, NM: Rio Grande Press, 1983): 174; Topeka *Daily Capital* (hereafter *Capital*), 28 January 1927.
11. Red Cloud quotation, Edwin L. Sabin, *Building the Pacific Railway* (Philadelphia: J. B. Lippincott Co., 1919): 233; Capital, 28 January 1927.
12. *Santa Fe Magazine* (hereafter *SFM*), January 1911, 35. I am very grateful to Mike Martin, Public Relations officer of the Santa Fe Corporate office in San Bernardino for providing me with access to their fine collection of the *Santa Fe Magazine*.
13. *Iron Cages: Race and Culture in Nineteenth-Century America* (Seattle: University of Washington Press, 1979): 193.
14. *Pacific Railway Commission*, 42.
15. Tzu-Kuei, "Chinese Workers," 53.

16. *Pacific Railway Commission*, 43.
17. Ping Chiu, *Chinese Labor in California, 1850–1880* (Madison: University of Wisconsin Press, 1963): 49.
18. Reprinted in the *Optic*, 28 January 1881.
19. *El Paso Daily Times* (hereafter *EPT*), 30 March 1905.
20. Benjamin Brooks, "The Southwest from a Locomotive," *Scribner's Magazine* (October 1903): 432; Reprinted in the *SFM*, September 1907, 370.
21. W. Elliot Brownlee, *Dynamics of Ascent: A History of the American Economy* (New York: Alfred A. Knopf, 1974) 101; Licht, *Working for the Railroad: The Organization of Work in the Nineteenth Century* (Princeton, New Jersey: Princeton University Press, 1983): 10; Takaki, *Iron Cages* (1979): 174.
22. John Coatsworth, *Growth Against Development: The Economic Impact of Railroads in Porfirian Mexico* (DeKalb: Northern Illinois University Press, 1981): 7; Chiu, *Chinese Labor*, 40.
23. James H. Ducker, *Men of the Steel Rails: Workers on the Atchison, Topeka & Santa Fe Railroad, 1869–1900* (Lincoln: University of Nebraska Press, 1983): xi.
24. Alfred D. Chandler, *The Visible Hand: The Managerial Revolution in American Business* (Cambridge: Harvard University Press, 1977): 204.
25. Licht, *Working for the Railroad*, 10.
26. *EPT*, 4 May 1900; also quoted in García, *Desert Immigrants*, 16.
27. García, *Desert Immigrants*, 20.
28. Meinig, *Southwest*, 41.
29. White, *It's Your Misfortune*, 258; *Arizona Weekly Star* (hereafter *Star*), 26 December 1878.
30. Gerald D. Nash, *The American West in the Twentieth Century: A Short History of an Urban Oasis* (Englewood Cliffs, New Jersey: Prentice Hall, 1973): 32; Rickard, *A History of American Mining* (1932); Helen Zeese Papanikolas, "Life and Labor Among the Immigrants of Bingham Canyon," *Utah Historical Quarterly* 33 (Fall 1965): 289–325; Katherine R. Elorreaga, Lisa M. Gallegos and Alberto L. Pulido, "Labor and Community Formation in Utah: The Mexican American Experience in Bingham Canyon," paper presented at the National Association for Chicano Studies, San Antonio, Texas, March 26–28, 1992.
31. Nash, *The American West*, 33, 32.

32. García, *Desert Immigrants*, 17.
33. Ibid., 18.
34. Ibid., 17.
35. Ibid., 19.
36. Ibid., 19–20.
37. Richard L. Nostrand, *The Hispano Homeland* (Norman: University of Oklahoma Press, 1992): 85, 116; Arthur L. Campa, *Hispanic Culture in the Southwest* (Norman: University of Oklahoma Press, 1993): 148.
38. *EPT*, 14 January 1902.
39. John Stover, *American Railroads* (Chicago: n.p., 1961): 98.
40. John T. Schlebecker, *Cattle Raising on the Plains, 1900–1961* (Lincoln: University of Nebraska Press, 1963): 21.
41. White, *It's Your Misfortune*, 222.
42. *Lone Star*, 28 November 1885; for consul quote see Charles W. Kindrick to W. R. Day, Assistant Secretary of State, Ciudad Judrez, January 12, 1898, General Records of the Department of State, Washington D.C., Record Group 59, Microfilm roll 6. Also cited in García, *Desert Immigrants*, 23.
43. Nash, *The American West*, 26; John F. Stover, *American Railroads* (Chicago: University of Chicago Press, 1961): 184.
44. White, *It's Your Misfortune*, 262.
45. *SFM*, February 1927, 21; White, *It's Your Misfortune*, 257.
46. *SFM*, February 27; see also *SFM*, December 1911, 51–54; *SFM*, April 1917, 107–110; *SFM*, January 1924, 51–53.
47. *SFM*, January 1924, 51; *SFM*, February 1927, 24.
48. *SFM*, February 1927, 24.
49. *San Bernardino Daily Sun* (hereafter *Sun*), 18 June 1909.
50. Robert J. Rosenbaum, *Mexicano Resistance in the Southwest: "The Sacred Right of Self-Preservation"* (Austin: University of Texas Press, 1981): 99–124.
51. John Coatsworth, *Growth Against Development: The Economic Impact of Railroads on Porfirian Mexico* (DeKalb: Northern Illinois University Press, 1981): 79.
52. Tzu-Kuei, "Chinese Workers," 53; Jim F. Heath, "A Study of the Influence of the Atchison, Topeka & Santa Fe Railroad Upon the Economy of New Mexico, 1878–1900" (M.A. thesis, University of New Mexico, 1955): 10; *Tombstone Epitaph* (hereafter *Epitaph*), 5 August 1881; Lorena M. Parlee, "The Impact of United States Railroad Unions on Organized Labor

and Government Policy in Mexico (1880–1911)," *Hispanic American Historical Review* 64 (August 1984): 446; *The Railroad Gazette*, 23 July; 26 November 1880.

53. *Arizona Weekly Star*, 5 February 1880.
54. Robert E. Quirk, *An Affair of Honor: Woodrow Wilson and the Occupation of Veracruz* (New York: Norton, 1962): passim.
55. Some important works are: John Kenneth Turner, *Barbarous Mexico* (Austin: University of Texas Press, [1910] 1969); Charles C. Cumberland, *Mexican Revolution: Genesis Under Madero* (Austin: University of Texas Press, 1952); John Womack, Jr., *Zapata and the Mexican Revolution* (New York: Vintage, 1969); James D. Cockroft, *Intellectual Precursors of the Mexican Revolution, 1900–1913* (Austin: University of Texas Press, 1971); Ramón Eduardo Ruiz, *The Great Rebellion: Mexico, 1905–1924* (New York: W. W. Norton, 1980); Alan Knight, *The Mexican Revolution*, 2 volumes (Lincoln: University of Nebraska Press, 1986); John Mason Hart, *Revolutionary Mexico: The Coming Process of the Mexican Revolution* (Berkeley: University of California Press, 1989).
56. John Coatsworth, "Railroads, Landholding, and Agrarian Protest in the Early Porfiriato," *Hispanic American Historical Review* 54 (February 1974): 48.
57. The Dillingham Commission also noted this transformation, see U.S. Congress, Senate, Dillingham Commission, *Immigrants in Industries* (Washington D.C.: Government Printing Office 1911) Part 111-1, 12 (hereafter Dillingham Commission). Reports on the issue of climate are pervasive. See Stone Report 86. *Railway Age Gazette* (hereafter *RAG*), 20 September 1912, 529; Railway Maintenance Engineer (hereafter *RME*) 13 March 1917, 71.
58. W. Thomas White, "A History of Railroad Workers in the Pacific Northwest, 1883–1934" (Ph.D. dissertation, University of Washington, 1981): 3.
59. Ibid., 263. Contrary to White's findings, railroads did recruit Mexicans into the Pacific Northwest as track laborers as my research shows.
60. See the *Union*, 13 May 1881; *Epitaph*, 25 May 1882; *Albuquerque Morning Journal*, 26 January; 8, 16 March; 4, 7, April 1907; *Sun*, 2 April 1909; *EPT*, 25 February 1902; 7, 8 January; 2 March 1907; 12 October 1910.
61. *Arizona Sentinel*, 23 November 1878.

62. Reported in the *Los Angeles Herald*, 14 January 1882.
63. *Epitaph*, 25 May 1882.
64. Ibid.
65. *Epitaph*, 7 April 1882.
66. SFM, June 1909, 749; *Santa Barbara Morning Press* (hereafter *Morning Press*) 8 June 1894; 29 August 1899; 22 May, 13 June 1900; Camarillo, *Chicanos in a Changing Society*, 137–38; for Mexican cooks see *Morning Press*, 2 May 1900.
67. "Santa Fe Splinters" (hereafter "Splinters") 1940, KSHS.
68. *Arizona Silver Belt*, 28 April 1898; Mary Roberts Coolidge, *Chinese Immigration* (New York: H. Holt & Company, 1909): 385.
69. *Los Angeles Socialist* (hereafter *Socialist*) 15 March 1902.
70. SFM, June 1909, 749.
71. Clark, "Mexican Labor," 1908, 466; also discussed in *Miners Magazine*, 31 December 1908, 13.
72. Coolidge, *Chinese Immigration*, 385.
73. Waters, *Steel Trails to Santa Fe*, 327.
74. Dillingham Commission, Part III–1. 12.
75. *Los Angeles Times* (hereafter LAT), 5 June 1925.
76. See Harry Braverman, *Labor and Monopoly Capital: The Degradation of Work in the Twentieth Century* (New York: Monthly Review Press, 1974): passim.
77. Charles H. Hufford, *The Social and Economic Effects of the Mexican Migration into Texas* (San Francisco: R&E Associates, [1929] 1971): 51.
78. RAG, 21 December 1900, also cited in Ducker, *Men of the Steel Rails*, 27; Vernon Monroe McCombs, *From over the Border, A Study of Mexicans in the United States* (New York: Council of Women for Home Missions and Missionary Education Movement, 1925): 24.
79. *The Indispensable Enemy: Labor and the Anti-Chinese Movement in California* (Berkeley: University of California Press, 1971): 66.
80. A contemporary discussion of Black track workers is provided in "The Negro as a Maintenance Laborer," RME, February 1917, 37–40.
81. Robert Athearn, *Rebel of the Rockies: The Denver and Rio Grande Western Railroad* (New Haven: Yale University Press, 1962): 56.

82. Carl Russell Graves, "Scientific Management and the Santa Fe Railway Shopmen of Topeka, Kansas, 1900–1925" (Ph.D. dissertation, Harvard University, 1980): 6.

Notes to Chapter 2

1. Clippings, KSHS Topeka.
2. *U.S. Bureau of the Census, Statistical Abstract of the United States*: 1982–83 (103rd edition) Washington, D.C., 1982.
3. See Clark, "Mexican Labor," 466; Mexicans in California (hereafter Mexicans in California), Report of C. C. Young's Mexican Fact-Finding Committee [1930] rpt., San Francisco: R&E Research Associates (1970): 90–91; Dillingham Commission, 8, 13; Stone Report, 80; Samuel Bryan, "Mexican Immigrants in the United States," *The Survey* (September 1912): 727–28; Paul S. Taylor, *A Spanish-Mexican Peasant Community: Arandas in Jalisco, Mexico* (Berkeley: University of California Press, 1933): 41; Union Pacific Payroll Records, Union Pacific Railroad Offices, Omaha, Nebraska, cited in Vicente V. Mayer, Jr., *Utah: A Hispanic History* (Salt Lake City: University of Utah, 1975): 39; *RME*, March 1920, 90–91; U.S. Bureau of Labor Statistics, "Increase of Mexican Labor," *Monthly Labor Review* 32 (January 1931): 85–86.
4. Leobardo F. Estrada, F. Chris García, Reynaldo Flores Macias and Lionel Maldonado, "Chicanos in the United States: A History of Exploitation and Resistance," *Daedalus* (Spring 1981): 113; García, *Desert Immigrants*, 35–37; see also Bryan, "Mexican Immigrants," 726.
5. Estrada, et al., "Chicanos in the United States," 113.
6. Sabin, *Building the Pacific Railway*, 172.
7. Vargas, *Proletarians*, 56–85.
8. B. A. Botkin and Alvin F. Harlow, *A Treasury of Railroad Folklore* (New York: Crown Publishers, 1953): 223; Melvyn Dubofsky, *We Shall Be All: A History of the IWW* (New York: Quadrangle, 1969): 313.
9. Meinig, *Southwest* (New York: Oxford University Press, 1971) passim; John R. Chavez, *The Lost Land: The Chicano Image of the Southwest* (Albuquerque: University of New Mexico Press, 1984): 9–10; Ramón A. Gutiérrez, *When Jesús Came, the*

Corn Mothers Went Away: Marriage, Sexuality, and Power in New Mexico, 1500–1846 (Stanford: Stanford University Press, 1991): 300–4.

10. García, *Desert Immigrants*, 15–16.
11. See Matt S. Meier and Feliciano Rivera, *The Chicanos: A History of Mexican Americans* (New York: Hill and Wang, 1972): 126; Laird (1975): 27; Mark Reisler, *By the Sweat of their Brow: Mexican Immigrant Labor in the United States, 1900–1940* (Westport: Greenwood Press, 1976): 3; Mines (1980): 26; AT&SF, Payroll Records, August 1874, KSHS; see also Virginia Pierce Hicks, "Sketches of Early Days in Kearny County," *Kansas Historical Quarterly* (February 1938): 55, mentions Mexican track workers in Kansas in 1872 but does not provide evidence. The exception is Sarah Deutsch, *No Separate Refuge: Culture, Class, and Gender on an Anglo-Hispanic Frontier in the American Southwest, 1880–1940* (New York: Oxford University Press, 1987): 19.
12. *Investigation of Labor Troubles in Missouri, Arkansas, Kansas, Texas, and Illinois* (Washington: Government Printing Office, 1887): hereafter *Labor Troubles*. Also cited in Theresa Ann Case, "Class as Culture: Gender and Race in the 1886 Southwest Strike," paper presented at the Southwest Labor Studies Conference, Santa Cruz, California, April 29, 1994. I am grateful to Margo Gutiérrez at the Benson Latin American Collection at the University of Texas at Austin for her assistance in locating this document.
13. *Star*, 7 November 1878.
14. Bryan, "Mexican Immigrants," 726–27.
15. For example, see Julio Norte, *Sun*, 15 February 1906; see also Fred García, "Indians Opposed Cholo Marriages," *Sun*, 19 February 1908; *SFM* December 1912, 99–100.
16. Cynthia Mines, "Riding the Rails to Kansas" (Masters thesis, McPherson College, 1980): 3, 26; Judith Ann Fincher Laird, "Argentine, Kansas: The Evolution of a Mexican-American Community, 1905–1940" (Ph.D. dissertation, University of Kansas, 1975): 116; "Outline of Study: Mexican Track Labor on the Atchison, Topeka and Santa Fe Railroad," Topeka, Kansas, Atchison, Topeka and Santa Fe Railway, General offices, August 1928, Typescript, Paul S. Taylor Collection, Bancroft Library (hereafter PTC). For Yaqui track workers see Immigrant Inspector

Frank R. Stone, Letter to Supervising Inspector F. W. Berkshire, Immigration Service, El Paso, Texas, 23 June 1910, Record Group 85 52546/31-B, 74, INS, National Archives. Hereafter referred to as Stone Report. See also W. R. Groshell, "How Mexicans are Employed," *RME*, March 1917, 72; see also "'Little Mexico' in Northern Cities," *World's Work*, September 1924, 46.

17. I thank Kurt M. Peters for allowing me to cite his work. See "Watering the Flower: The Laguna Pueblo-Santa Fe Railroad Employment Contract, 1880–1945," paper presented at the South West Labor Studies Association Annual Conference, Santa Cruz, California, 29 April 1994.
18. "Report of the Governor of New Mexico for the Year Ending June 30, 1898," *Annual Reports of the Department of the Interior for the Fiscal Year End, June 30, 1898*. House of Representatives, 55th Congress, 3d. Sess. Document no. 5. p. 561; "Leupp Wants Indians to Work," *EPT*, 23 December 1906; "Will Employ Many Indians," *Sun*, 29 March 1906; "Indians Employed by the Santa Fe," *Albuquerque Evening Citizen*, 28 March 1906.
19. *San Diego Union*, 3 June 1881; Ronald Takaki, *Iron Cages: Race and Culture in 19th-Century America* (Seattle: University of Washington Press, 1979): 193.
20. Kurt M. Peters, "Watering the Flower: The Laguna Pueblo and the Santa Fe Railroad, 1880–1943," in *Native Americans and Wage Labor: Ethnohistorical Perspectives*, ed. Alice Littlefield and Martha C. Knack (Norman: University of Oklahoma Press, 1996).
21. Allan Harper, Andrew Cordova, and Kalervo Oberg, *Man and Resources in the Middle Rio Grande Valley* (Albuquerque: University of New Mexico Press, 1943): 64; see also Joseph F. Park, "The History of Mexican Labor in Arizona During the Territorial Period" (M.A. thesis, University of Arizona, 1961): 229; Clark, "Mexican Labor," 494.
22. Deutsch, *No Separate Refuge*, 19.
23. Elfido López Sr. Collection, Colorado Historical Society. I am grateful to Rebecca Lintz at the CHS for her kind assistance. Also cited in Deutsch (1987): 18.
24. James P. Craig, "How to Handle Mexicans," *SFM*, November 1914, 28; and later reprinted in *RME*, March 1917, 73.
25. Campa, *Hispanic Culture in the Southwest*, 116–17.

26. Nostrand, *Hispano Homeland*, 140; Frances Leon Swadesh, *Los Primeros Pobladores: Hispanic Americans of the Ute Frontier* (Notre Dame: University of Notre Dame Press, 1974): 98; Ducker, *Men of the Steel Rails*, 27.
27. James Lewellyn Allhands, *Railroads to the Rio* (Salado, Texas: Anson Jones Press, 1960): 57.
28. *The Iron Road*, 60 min., distributed by PBS video, Alexandria, Virginia, 1990.
29. García, *Desert Immigrants*, 204.
30. Charles F. Lummis, *The Land of Poco Tiempo* (New York: Charles Scribners and Sons, 1925): 236–37.
31. Camarillo, *Chicanos in a Changing Society*, 137–38; Elfido López Sr. Collection, Colorado Historical Society.
32. Herbert G. Gutman, *Power and Culture: Essays on the American Working Class* (New York: The New Press, 1987): 255–59.
33. Gerd Korman has examined this compelling link between industrialization and Americanization in Milwaukee. See his book, *Industrialization, Immigrants, and Americanizers: The View from Milwaukee, 1866–1921* (Madison: State Historical Society of Wisconsin, 1967).
34. I borrow this analysis from Elizabeth Jameson, "Women as Workers, Women as Civilizers: True Womanhood in the American West," in *The Women's West*, ed. Susan Armitage and Elizabeth Jameson (Norman: University of Oklahoma,Press, 1987): 145–64.
35. Clark, "Mexican Labor," 466, 521; *RAG*, 20 September 1912, 529; Reisler, *By the Sweat of their Brow*, 13; Laird, "Argentine," 118; García, *Desert Immigrants*, 38.
36. Perhaps the best known are the foundational writings of Carey McWilliams, *North From Mexico: The Spanish Speaking People of the United States* (New York: Greenwood Press, 1968); and Octavio I. Romano V., "Minorities, History, and the Cultural Mystique," *El Grito* (Fall 1967): 5–11; "The Anthropology and Sociology of the Mexican-Americans: The Distortion of Mexican-American History," *El Grito* (Fall 1968): 13–26; "The Historical and Intellectual Presence of Mexican Americans," *El Grito* (Winter 1969): 32–46; "Social Science, Objectivity and the Chicanos," *El Grito* (Fall 1970): 30–42.
37. López interview, 22 October 1990.
38. Aparicio interview, 23 December 1990.

39. Interview with John and Pascual Madrigal. April 29, 1980. Interviewed by Laurie Bretz, KSHS, my translation.
40. *Tetwiler's*, 13 June 1907.
41. Vargas, *Proletarians*, 37.
42. El Circulo de Obreros Católicos Mexicanos, Indiana Harbor, Indiana, PTC, 192.
43. Anthony Quinn, *The Original Sin: A Self-Portrait* (Boston: Little, Brown and Company, 1972): 52.
44. Ibid.
45. Ibid.
46. Rutter, "Mexican Americans," 86; García, *Desert Immigrants*, 40, fn. 17; cites Judith Laird who cites Rutter. Vargas has found similar evidence, see *Proletarians*, 37.
47. For instance see, Juan Gómez-Quiñones, "Mexican Immigration to the United States and the Internationalization of Labor, 1848–1980: An Overview," in *Mexican Immigrant Workers in the U.S.*, ed. Antonio Rios Bustamante (Los Angeles: Chicano Studies Research Center Publications, 1981): 13–34.
48. In English the term *peon* laborer referred specifically to those who worked at any variety of tasks (including track work) on large agricultural estates called Haciendas. Later the term *cholo* is used frequently in the documents to refer to the Mexican lumpenproletariat.
49. For example, see J. D. Mathews, "The Mexican as Track Laborer"; L. J. Hughes, "Good Treatment Necessary for Mexicans"; M. Ganley, "Characteristics of the Mexican"; Resident Engineer, "Pecularities of Mexican labor," *RAG*, 20 September 1912, 527–30.
50. Hufford, *Mexican Migration*, 37.
51. "Splinters," KSHS; see also García, *Desert Immigrants*, 35.
52. Clark, "Mexican Labor," 466.
53. Bryan, "Mexican Immigrants," 727.
54. AT&SF Payroll Records, KSHS.
55. The literature on Mexican immigration is extensive. A few representative works are, Manuel Gamio, *Mexican Immigration to the United States: A Study of Human Migration and Adjustment* (1930; rpt., New York: Dover, 1971); Paul S. Taylor, 7 volumes, *Mexican Labor in the United States* (1929–1934); Mark Reisler, *By the Sweat of their Brow: Mexican Immigrant Labor in the United States, 1900–1940* (Westport, Connecticut:

Greenwood Press, 1976); García, *Desert Immigrants*; Camille Guerin-Gonzales, *Mexican Workers and American Dreams: Immigration, Repatriation, and California Farm Labor, 1900–1939* (New Brunswick: Rutgers University Press, 1994).

56. U.S. Industrial Commission, *Reports of the Industrial Commission on Immigration*, vol. 15, 1901, 749, rpt. Arno Press 1970 (hereafter Industrial Commission).
57. Dillingham Commission, pt. 25, vol. III, pt. I, p. 7. Also cited in García, *Desert Immigrants*, 40.
58. Bryan, "Mexican Immigrants," 727–28.
59. Clark, Bulletin 78 466.
60. McWilliams, *North From Mexico* (1968): 156; Barrera, *Race and Class in the Southwest* (1979): 84–85; Estrada, et al., "Chicanos in the United States" (1981): 112; Colin M. MacLachlan, *Anarchism and the Mexican Revolution: The Political Trials of Ricardo Flores Magón in the United States* (Berkeley: University of California Press, 1991): 12.
61. Hufford, *Mexican Migration*, 51.
62. *Mexicans in California*, 71, 91; also cited in Wollenberg, "Working on El Traque," 367.
63. Hufford, *Mexican Migration*, 51.
64. Chicago Notes, PTC, 75, 72.
65. "Retrenchment in the Railway Labor Force in 1911," *RAG*, 1 September 1911, 405. This data represents only 60 percent of the returns.
66. "Increase of Mexican Labor," 85.
67. W. W. Pannell, "Mexican Workers in the Southwest," *International Socialist Review*, September 1916, 169.
68. Parlee, "Railroad Unions," 443–75.

notes to chapter 3

1. Gamio, *Mexican Immigration*, 85. The term *enganchado* is translated by Gamio to mean "the hooked ones" i.e., in contract labor situation. I also correct Gamio's translation of "Nos pusimus a trampar" to "tampar" which is a more logical reading of the corrido. "Trampear" means "to tramp" (travel on the rails as an itinerant laborer) and "tampar" is how traqueros referred to "tamping," as in tamping the roadbed to secure the ties.

2. Ernesto Galarza, "Mexicans in the Southwest: A Culture in Process," in *Plural Society in the Southwest*, Edward Spicer and Raymond H. Thompson, eds. (Albuquerque: University of New Mexico Press, 1975): 279.
3. See Montgomery, *The Fall of the House of Labor*, 58–111; and Andrea Graziosi, "Common Laborers," Labor History 22 (Fall 1981): 512–544; and more recently the work of Peter Way has received much deserved attention; see his book *Common Labour: Workers and the Digging of the North American Canals, 1780–1860* (Cambridge: Cambridge University Press, 1993), passim.
4. Graves, "Scientific Management," 182.
5. Montgomery, *Fall of the House of Labor*, 58.
6. Graves, "Scientific Management," 183; Robert A. Swan, "The Ethnic Heritage of Topeka: Immigrant Beginnings" (Topeka: Institute of Comparative Ethnic Studies, (1974): 124–56; AT&SF Payroll Records, KSHS.
7. *Railway Engineering and Maintenance of Way* (hereafter *REMW*), December 1906, 270–71.
8. E. E. Russell Tratman, *Railway Track and Track Work* (New York: McGraw-Hill, 1909): 309; *Los Angeles Daily Herald*, 15 January 1882.
9. Ibid.
10. López interview.
11. *RAG*, March 1910, 559; *SFM*, October 1909, 1326; June 1923, 68; Manuel Preciado, foreman on the AT&SF line from National City to Colton, Splinters, KSHS; Dillingham Commission, 111-1, 23. See also Gary W. Kennedy to J. J. Davis, Secretary of Labor, December 22, 1922, Federal Mediation and Conciliation Service, National Archives, RG 28011 files 165/223B, Box 13.
12. Least Heat-Moon, *PrairyErth*, 233–34.
13. Ibid.
14. *Los Angeles Record* (hereafter *Record*), 9 March 1903.
15. Ibid.
16. *Sun*, 4 September 1909.
17. Stover, *American Railroads*, 151–52; also cited in Licht, *Working for the Railroad*, 184.
18. *SFM*, February 1914, 63–65.
19. James Lewellyn Allhands, *Gringo Builders* (Iowa City: Privately printed, 1931) 166.

20. The Dillingham Commission (1911) reported more than 50 percent of Mexican wives accompanied their husbands to find work in the United States. See Elizabeth Salas, *Soldaderas in the Mexican Military: Myth and History* (Austin: University of Texas Press, 1990).
21. Quinn, *Original Sin*, 54–55.
22. Ibid.
23. Ibid., 56.
24. Ibid., 58.
25. Graves, "Scientific Management," 45–46.
26. *REMW*, October 1915, 318.
27. Graves, "Scientific Management," 182; Ducker, *Men of the Steel Rails*, 4; Licht, *Working for the Railroad*, 224.
28. Ducker, *Men of the Steel Rails*, 4.
29. *Engineering News and American Railway Journal*, 15 June 1893, 564.
30. Clark, "Mexican Labor," 496; Vargas, *Proletarians*, 89.
31. Clark, "Mexican Labor," 485, 496; Dillingham Commission, pt. 25, Vol. III, pt. I, pp. 7, 13–14, 19, 23; Edwin F. Bamford, "Industrialization and the Mexican Casual," *Proceedings of the Southwestern Political and Social Science Association* (Austin: March 1924): 255; Wollenberg, "Working on El Traque," 362.
32. *RME*, March 1917, 72.
33. Ibid., 73.
34. Chicago Notes, PTC, 89.
35. *SFM*, January 1913, 86.
36. Ibid.
37. *Albuquerque Morning Journal*, 21 March 1907.
38. *Tombstone Epitaph*, 10 August 1881; see also Arnoldo De Leon, *They Called Them Greasers: Anglo Attitudes Toward Mexicans in Texas, 1821–1900* (Austin: University of Texas Press, 1983).
39. *SFM*, November 1914, 27.
40. Hufford, *Mexican Migration*, 37.
41. *SFM*, November 1914, 28; see also Bamford, "Mexican Casual," 255.
42. *RME*, March 1917, 71.
43. *RAG*, 20 September 1912; reprinted in *SFM*, January 1913, 86.
44. *SFM*, November 1914, 28; reprinted and edited in *RME*, March 1917, 72 3.
45. *SFM*, June 1909, 745.

46. *RAG*, 20 September 1912, 529; see also *SFM*, June 1909, 749.
47. *RME*, March 1917 71.
48. *RME*, March 1917, 73–74.
49. William Z. Foster, *Pages from a Worker's Life* (New York: International Publishers, [1939] 1978): 28–29.
50. *RAG*, 20 September 1922, 529.
51. *SFM*, October 1912, 64.
52. *RAG*, 20 September 1912, 529.
53. Taylor, Arandas, 41; see also Bryan, "Mexican Immigrants," 728; Wollenberg, "Working on El Traque," 365.
54. *RAG*, September 1912, 529; *SFM*, November 1914, 29; *RME*, March 1917, 73. Also meaning to "kill time or to shirk," *REMW*, December 1911, 556.
55. *RAG*, September 1912, 529.
56. Ibid.
57. *RME*, March 1917, 72.
58. *RAG*, September 1912, 529; SFK, January 1913, 86.
59. Aparicio interview, 23 December 1990.
60. Circulo de Obreros Católicos Mexicanos, Indiana Harbor, Indiana. PTC.
61. Vargas, *Proletarians*, 46, 93.
62. *RAG*, September 1912, 528.
63. Ibid., 529.
64. *SFM*, November 1914, 27.
65. Ibid. 165
66. Ibid.
67. Ibid.; *EPT*, 25 September 1910; 13 October 1910; *El Cosmopolita*, 29 July 1916; 14 July 1917. *Patroncito* might also be translated as "middle-man," "interpreter," or "foreman's assistant."
68. *SFM*, January 1914, 41.
69. *SFM*, November 1914, 27.
70. Ibid.
71. Ibid.
72. Daniel Venegas, *Las Aventuras de Don Chipote: O Cuando Los Pericos Mamen* (Mexico: Secretaria de Educación Publica, [1928] 1984) passim.
73. *SFM*, November 1914, 29.
74. Carlos Fuentes, "Hispanic USA: The Mirror of the Other," *The Nation*, 30 March 1992, 409. Ramírez died in 1960.

75. *RAG*, September 1912, 530.
76. Ibid.
77. *SFM*, November 1914, 27.
78. *Tetwiler's*, 21 February 1907.
79. The origin of the term *chante* is unclear, but there seems to be a convergence of both Nahuátl and English. "Shanty" is the term Irish workers used to describe a small crude shack. It is plausible that traqueros' linguistic difficulty with the "sh" sound pronounced "shanty" as "*chante*." For an example of this pronunciation see William Least Heat-Moon, *PrairyErth* (Boston: Houghton Mifflin Company, 1991): 231. On the other hand, Guillermo Lux and Maurilio E. Vigil state that the term derives from Nahuátl. See their essay, "Return to Aztlan: The Chicano Rediscovers His Indian Past," in *Aztlan: Essays on the Chicano, Homeland*, Rodolfo Anaya and Francisco Lomeli, eds. (Albuquerque: University of New Mexico Press, 1989): 93–110; Lux and Vigil cite Jorge Thomas, unpublished study "Influencia Azteca en la cultura y lenguaje de Nuevo Mejicoll (M.A. thesis, New Mexico Highlands University, 1974): 49–54. According to James Lockhart, *The Nahuas After the Conquest: A Social and Cultural History of the Indians of Central Mexico, Sixteenth Through Eighteenth Centuries* (Stanford: Stanford University Press, 1992): 60. The term *chan* was Nahuátl for home, and occasionally the *-tli* was attached to it as in *chan-tli*. Over the years, this term may have been truncated to *chante*. In either case the term "chante" does not carry with it the same pejorative connotations as the term "shanty." See Gamio, *Mexican Immigration*, 232. By the 1940s *pachucos* and urban streetwise Chicanos adopted the term for "home." I thank Rebecca Horn for sharing her knowledge of Nahuátl with me.
80. *SFM*, June 1923, 42.
81. Manuel Gamio, *The Life Story of the Mexican Immigrant* (New York: Dover Publications Inc., 1971): 150.
82. John Generella, Employment Agent and Commissary for Chicago and North Western Railway, Chicago, July 11, 1928, PTC.
83. Luis García, South Chicago, June 7, 1928, PTC.
84. F. Huerta, Chicago, June 8, 1928, PTC.
85. *RME*, September 1920, 337; see also *RME*, February 1921, 47; *RME*, May 1929, 45.
86. See Vargas, *Proletarians*, passim.

87. Rexford G. Tugwell, "The Casual of the Woods," *The Survey*, July 1920, 473.
88. Wisconsin Industrial Commission Report, Labor Camps in Wisconsin, 4 September 1914, 20. Although this report makes no reference to the presence of Mexicans by name, I presume that Mexicans are present. Nonetheless, the conditions described are instructive.
89. *Railway Employee's Journal*, 30 July 1903; see also Valiant, Wisconsin Industrial Commission, 20.
90. Wisconsin Industrial Commission, 21. 167
91. Recent studies show the variation of homosexual desire and identity in Mexican culture, see Tomas Almaguer, "Chicano Men: A Cartography of Homosexual Identity and Behavior," *Differences: A Journal of Feminist Cultural Studies* 3 (1991): 75–100.
92. Tugwell, "Casual of the Woods," 473.
93. Ibid.
94. Carlton H. Parker, "The California Casual and His Revolt," *Quarterly Journal of Economics* (November 1915): 118.
95. Ibid.
96. Ibid., 124.
97. *Record*, 6 April 1903; 13 April 1903.
98. *EPT*, 24 January 1909.

Notes to Chapter 4

1. *LAT*, 26 April 1903. *The Los Angeles Times* reported only thirty protesting mexicanas, but given its heavy anti-union and racist bias it is likely that at least twice the number of mexicanas participated.
2. Ibid.
3. Lorena Parlee, "The Impact of United States Railroad Unions on Organized Labor and Government Policy in Mexico (1880–1911)," *Hispanic American Historical Review* (August 1984): 443–75; this article is drawn from her fine dissertation, "Porfirio Diaz, Railroads, and Development in Northern Mexico: A Study of Government Policy Toward the Central and Nacional Railroads, 1876–1910" (Ph.D. dissertation,. University of California, San Diego, 1981).

4. Parlee (1984): passim.
5. *Solidarity*, 22 July 1916; D. W. Hertel, *History of the Brotherhood of Maintenance of Way Employees: Its Birth and Growth, 1887–1955* (Washington D.C.: Ransdell Inc., 1955): 3; Reisler (1976): 48; Graves, "Scientific Management," 48.
6. The National Protective Association of Locomotive Engineers was the first railroad union founded in 1855. See Licht, *Working for the Railroad*, 240.
7. Andrew Bancroft Schlesinger, "Las Gorras Blancas, 1889–1891," *Journal of Mexican American History* (Spring 1971): 87–143. Rosenbaum, *Mexicano Resistance*, 99; Deutsch, *No Separate Refuge* (1987): 25–26.
8. Nostrand, *Hispano Homeland*, 116–17.
9. Schlesinger, "Las Gorras Blancas," 95; Rosenbaum, *Mexicano Resistance*, 98–110; Rodolfo Acuna, *Occupied America* (New York: Harper and Row, 1988): 71; Roxanne Dunbar-Ortiz, *Roots of Resistance: Land Tenure in New Mexico, 1680–1980* (Los Angeles: Chicano Studies Research Center Publications, 1980): 91–109.
10. Howard R. Lamar, *The Far Southwest, 1846–1912: A Territorial History* (New Haven: Yale University Press, 1966): 139; Clark S. Knowlton, "Land-Grant Problems Among the State's Spanish-Americans," *New Mexico Business* (June 1967): 3–4.
11. For examples of nineteenth century secret societies see Leon Fink, *Workingmen's Democracy: The Knights of Labor and American Politics* (Urbana: University of Illinois Press, 1983); and Anthony Bimba, *The Molly Maguires* (New York Publishers, 1970).
12. Ibid., 121.
13. Ibid., 108–9; White Caps Investigation, L. B. Prince Papers, 1892, New Mexico State Records Center. I am grateful to J. Richard Salazar for his assistance in accessing these documents.
14. Letter to H. S. Waldo from C. D., Las Vegas, August 11, 1890.
15. White Caps Investigation.
16. Rosenbaum, *Mexicano Resistance*, 131.
17. White Caps Investigation.
18. Ibid.; depending on the circumstances, Anglos made little distinction between "Mexicans" and so-called Spanish residents.
19. Ibid.
20. Ibid. *Serbidores* could also mean "guardians."

21. Ibid.
22. Ibid.
23. Letter from Miguel Salazar to L. Bradford Prince, July 23, 1890.
24. Ibid.
25. *Optic*, 7 March 1890.
26. White Caps Investigation.
27. Payroll Records, AT&SF, KSHS; Jonathan Garlock, *Guide to the Local Assemblies of the Knights of Labor* (Westport: Greenwood Press, 1982): 295–96.
28. Ibid.; Lamar, *The Far Southwest*, 194.
29. Deutsch, *No Separate Refuge*, 26.
30. Fredrick C. Sanchez, "A History of the S.P.M.D.T.U," *San Luis Valley Historian* (Winter 1971): 3; see also José Timoteo López, *La Historia de la Sociedad Proteccion de Trabajadores Unidos* (New York: Comet Books, 1958): 7.
31. Ibid.
32. Deutsch, *No Separate Refuge*, 26.
33. Sanchez, "S.P.M.D.T.U," 1.
34. Jeremy Brecher, *Strike!* (Boston: Southend Press, 1987): 1–24.
35. Herbert Gutman, "Trouble on the Railroads in 1873–1874: Prelude to the 1877 Crisis," *Labor History* (Spring 1961): 217–19.
36. Case, "Class as Culture," 4, 8; see also Ruth A. Allen, *The Great Southwestern Strike* (Austin: University of Texas Publication, 1942): 21.
37. F. Huerta, Chicago, June 8, 1928, PTC.
38. *Railway Engineering Journal*, 23 July 1903; *Solidarity*, 12 August 1916; *Report of the Mayor's Commission on Unemployment* (Chicago: State Printing Office, March 1914): 70.
39. *Sun*, 6, 23, March 1908; *Tetwiler's*, 12 March 1908.
40. *Sun*, 27 March 1908.
41. Jeffrey M. Garcilazo, "Mexican Strike Activity in the Riverside and San Bernardino Areas, 1917," presented at the National Association for Chicano Studies, Sacramento, California, March 25, 1985.
42. *Sun*, 8 April 1917.
43. Ricardo Romo, *East Los Angeles: History of a Barrio* (Austin: University of Texas Press, 1983): 89–111; and Garcilazo, "Mexican Strike Activity," 20.

44. Ibid.; for citrus work and community see Gilbert G. González, *Labor and Community: Mexican Citrus Worker Villages in a Southern California County, 1900–1950* (Urbana: University of Illinois Press, 1994).
45. *SFM*, November 1914, 27; *RAG*, September 1912, 528, 30.
46. Zamora, *World of the Mexican Worker*, 69.
47. Ibid., 60.
48. Ibid., 60–61.
49. *LAT*, 26 April 1903.
50. Ricardo Romo, *East Los Angeles*, 29; George J. Sánchez, *Becoming Mexican American: Ethnicity, Culture and Identity in Chicano Los Angeles, 1900–1945* (New York: Oxford University Press, 1993): 88.
51. Pedro G. Castillo, "Making of a Mexican Barrio: Los Angeles, 1890–1920" (Ph.D. dissertation, University of California, Santa Barbara, 1979): 20.
52. *Record*, 9, 11, 12, March 1903.
53. *Record*, 24 April 1903. *The Record* and the *Times* refer to Nieto as A. N. or A. M. Nieto.
54. *LAT*, 25 March 1903.
55. Grace Heilman Stimson, *Rise of the Labor Movement in Los Angeles* (Berkeley: University of California Press, 1955): 267; Charles Wollenberg, "Working on El Traque: The Pacific Electric Strike of 1903," *California Historical Quarterly* 42 (August 1973): 362, this study remains as the only published study that focuses on the 1903 strike; Edward J. Escobar, "Chicano Protest and the Law: Law Enforcement Responses to Chicano Activism in Los Angeles, 1850–1936" (Ph.D. dissertation, University of California, Riverside, 1983): 205.
56. *Record*, 25 April 1903; also cited in Stimson (1955): 267; Wollenberg, (1973): 362; Escobar, (1983): 204–5.
57. *LAT*, 25 April 1903.
58. Ibid.
59. Ibid.; see Escobar (1983): 206–12.
60. *LAT*, 25 April 1903.
61. *Record*, 24 April 1903.
62. *LAT*, 28 April 1903.
63. *Record*, 2 May 1903.
64. *LAT*, 26 April 1903.
65. *LAT*, 27 April 1903.

66. *LAT*, 26 April 1903.
67. Ibid.
68. *Record*, 28 April 1903.
69. Ibid.
70. Ibid.; *Record*, 1 May 1903.
71. *LAT*, 10 May 1903; *Record*, 6, 7, 8, 11 May 1903.
72. *LAT*, 27 April 1903; *Record*, 28 April 1903; also cited in Stimson (1955): 267; Wollenberg (1973): 361.
73. *Record*, 2, 3 March 1904; also cited in Stimson (1955): 267; and Wollenberg (1973): 367.
74. *La Voz del Pueblo*, 6 September 1890; *EPT*, 29 November 1901; *Santa Fe New Mexican*, 2 June 1904; *EPT*, 18 April 1905, 16 July 1910; *Sun*, 8, 9, 17 April 1917; *Southern California Labor Press*, 26 September 1925; Licht notes an early agreement among white street railway workers in St. Louis, Missouri, in April 1902; see *Working for the Railroad*, 268.
75. *Record*, 16 March 1903.
76. *Socialist*, 2 May 1903; also cited in Wollenberg (1973): 366.
77. See James A. Sandos, *Rebellion in the Borderlands* (Norman: University of Oklahoma Press, 1992): Ward S. Albro, *Always a Rebel* (Fort Worth: Texas Christian University Press, 1992); Colin MacLachlan, *Anarchism and the Mexican Revolution* (Berkeley: University of California Press, 1991): W. Dirk Raat, *Revoltosos* (College Station: Texas A&M University Press, 1981): Juan Gomez-Quiñones, *Sembradores* (Los Angeles: Chicano Studies Center Publications, 1977): Lowell Blaisdell, *The Desert Revolution* (Madison: University of Wisconsin Press, 1962).
78. Chicago Notes, PTC.
79. Arnoldo DeLeon, *They Called Them Greasers: Anglo Attitudes Toward Mexicans in Texas, 1821–1900* (Austin, University of Texas Press, 1983).
80. Chicago Notes, McCourtney, Chief Clerk, Illinois Central, June 1928, PTC.
81. *SFM*, June 1909, 748.
82. Marston, "Mexican Traits," *The Survey*, August 1920, 563.
83. Letter O. A. Morse to A. J. Fink, Confederación de Sociedades Mexicanos, Chicago, August 5, 1925. Chicago Notes PTC.
84. Ibid.
85. *SFM*, November 1914, 28.
86. Ibid.

87. See chapter 3 for discussion of work slow-downs: *SFM*, June 1909, 745.
88. Ibid.
89. Deutsch, *No Separate Refuge*, 19.
90. Kenneth L. Roberts, "Wets and Other Mexicans," *Saturday Evening Post*, 4 February 1928, 146.
91. Ibid., 11.
92. *Tetwiler's*, 21 December 1905.
93. Ibid.
94. *La Voz del Pueblo*, 18 October 1890, 4.
95. Roberts, "Wets and Other Mexicans," 142.
96. Rodrigo's Notes, June 18, 1928, PTC, 127.
97. *SFM*, December 1912, 130.
98. *SFM*, November 1924, 100.
99. Max Sylvannus Handman, "The Mexican Immigrant in Texas," *Political and Social Science Quarterly* (June 1926): 37.
100. A useful book is Richard Graham, ed., *The Idea of Race in Latin America, 1870–1940* (Austin: University of Texas Press, 1990).
101. Thomas McKenney, May 31, 1928, Illinois Steel Company, PTC.
102. Ibid.
103. Chicago Notes, July 12, 1928, PTC.
104. Ibid.
105. *RAG*, 15 March 1910, 559; *SFM*, July 1931, 65; for a discussion of race and American workers see David R. Roediger, *Wages of Whiteness: Race and the Making of the American Working-Class* (London: Verso, 1991).
106. Ramon F. Adams, *The Language of the Railroader* (Norman: University of Oklahoma Press, 1977): 66. The term refers to tools used by track workers made by the Gandy Manufacturing Company in Chicago.

notes to chapter 5

1. García, *Desert Immigrants*, 50.
2. Given the variety of experiences, and the paucity of early source material, it is impossible to determine exactly where the first Mexican boxcar community appeared.

3. See Camarillo, *Chicanos in a Changing Society*, 53–78.
4. See García, *Desert Immigrants*, 34–36.
5. Licht, *Working for the Railroad*, 142; See also David L. Lightner, "Construction on the Illinois Central Railroad," *Monthly Labor Review* (June 1973): 290; for an outdated yet useful study of railroad worker camps in Canada see Edwin W. Bradwin, *The Bunkhouse Man: A Study of Work and Pay in the Camps of Canada, 1903–1914* (New York: Columbia University Press, 1928): passim.
6. William Abraham Bell, *New Tracks in North America* (London: Chapman and Hall, 1870): 90; also cited in the *SFM*, June 1907, 221; See also Atchison, Topeka and Santa Fe Payroll Records, KSHS.
7. Mr. and Mrs. James Rose Harvey, "Engineer Walk of the Denver & Rio Grande," *Colorado Magazine*, November 1947, 246; also cited in Frederick C. Sánchez, "A History of the S.P.M.D.T.U," *San Luis Valley Historian* (Winter 1971): 2.
8. Ibid.
9. *Optic*, 23 November, 1881.
10. Letter from A. A. Robinson to Allen Manvel, AT&SF, Minutes, KSHS.
11. "Causes of Mexican Migration to the United States: A Summary View," *Perspectives in American History* 7 (1973): 560.
12. See L. C. Lawton, "Mexican Laborer's Houses on the Santa Fe," *RAG*, August 1911, 344; Mr. Donahue, Roadmaster, Chicago Division, Chicago & Alton Railroad, July 1928. Chicago Notes, PTC.
13. Allhands, *Gringo Builders*, 258.
14. Ibid.
15. *EPT*, 11 January 1906.
16. Ibid.
17. "Meade Manuscript," Santa Fe Executive Files, KSHS, 145; Minnie M. Miller, "Heritage of Mexican-Americans Is Traced," *Emporia Gazette*, 10 September 1979.
18. See Graves, "Scientific Management," 1980; "To Provide for Employees," *Santa Fe New Mexican*, 19 August 1904.
19. Lawton, *RAG*, 18 August 1911, 344; "Erecting Mexican Laborer's Houses," *SFM*, September 1911, 75–76.
20. *RME*, March 1917, 71.
21. Mr. Donahue, Roadmaster, Chicago Division, Chicago & Alton

Railroad, July 1928, PTC.

22. *RAG*, 20 September 1912, 527; 528; *SFM*, January 1913, *RME*, March 1917, 72.
23. *RME*, March 1917, 71.
24. This was particularly important after the IWW was founded in 1905, since the IWW unlike the AFL, championed the "hobo" or "casual laborer."
25. E. L. M., "The Wife's Part in Railroading," *SFM*, April 1913, 70.
26. *RME*, March 1917, 73.
27. Ibid.
28. Mr. Courtney, Chief Clerk to Engineer in charge of Maintenance, Chicago, Illinois Central, June 1928. Chicago Notes, PTC.
29. Mexican laborers were also recruited as far east as New York, but there is no available evidence of Mexican boxcar communities there. George Rivera, Jr. and Juventino Mejía, "Mas aya del ancho rio: Mexicanos in Western New York," *Aztlan* 7 (Fall 1976): 499–513.
30. *Socialist*, 15 March 1902.
31. Ibid.
32. Ibid.
33. Ibid.
34. Stone Report, 75.
35. Ibid.
36. Ibid.
37. *RAG*, August 1911, 344; and *SFM*, September 1911, 75–76.
38. *RAG*, August 1911, 344.
39. Stone Report, 87; *SFM*, March 1921, 76.
40. "Mexicans in Kansas," Clippings, KSHS.
41. *SFM*, March 1921, 76; "Mexicans in Kansas," Clippings, KSHS; and *Emporia Gazette*, 10 September 1979.
42. William Least Heat-Moon, *PrairyErth* (Boston: Houghton Mifflin Company, 1991): 232.
43. Ibid.
44. López interview.
45. Mrs. Kel Gershom Center, July 17, 1928, Conversations with, and Regarding Mexicans, PTC.
46. Stone Report, 87.
47. Proviso Camp, June 24, 1928, Rodrigo's Notes, PTC.
48. López interview.

49. Rodrigo's Notes, PTC.
50. Ibid.
51. Ibid.
52. R. W. Nicol, Elgin, Joliet & Eastern Railroad, July 11, 1928, and Chicago Notes, 95, PTC.
53. *REMW*, March 1928, 111.
54. Nicol, PTC.
55. Robert N. McLean, "Rubbing Shoulders on the Border," *The Survey*, May 1924, 201.
56. Interview with José and Esperanza Cuevas, Topeka, Kansas, 18 October 1990.
57. *SFM*, August 1910, 95.
58. Interview with Antonia Aparicio, San Bernardino, California, November 24, 1990.
59. García, *Desert Immigrants*, 138.
60. See Mines, "Riding the Rails," 84; Miller, 10 September 1979.
61. Joe Gandelman, "Hispanic Identity: Tradition, Pride, vs. 'Melting Pot'," *Wichita Eagle*, 29 September 1980.
62. Zaragosa Vargas, "Life and Community in the 'Wonderful City of the Magic Motor': Mexican Immigrants in 1920s Detroit," *Michigan Historical Review* 15 (Spring 1989): 51.
63. Chicago Notes, PTC, 52.
64. Ottilio Falcone, June 18, 1928, Rodrigo's Notes, Chicago, Illinois, PTC. John Culhane, "Hero Street USA," *Reader's Digest*, May 1985, 81. The Mexican American community of Silvis, Illinois, was recently the focus of a documentary on a Public Television Station that depicted the life experiences of Mexican immigrants and their children from the teens through World War II.
65. *Sun*, 1 November 1906; 5 February 1908; *Tetwiler's*, 22 March 1906; 11 May 1905; *SFM*, December 1913, 88, 92; Pannell, "Mexican Workers," 169; see also Swan, "Ethnic Heritage," 144; and Harry H. Stoops and Lee Haney, "Ethnology," U.S. Work Projects Administration, *Historical Records Survey, Missouri* (April 27, 1936): 3.
66. *Tetwiler's*, 21 December 1905.
67. *Sun*, 5 June 1909.
68. *Sun*, 26 December 1909.
69. *RME*, March 1917, 71.
70. Paul T. Porter, "Handling Mexican Labor in California,"

Electric Railway Journal, 11 August 1928, 222; *Pacific Electric Magazine*, 28 July 1928.

71. *Electric Railway Journal*, 11 August 1928, 222.
72. *Pacific Electric Magazine*, 10 June 1926, 7.
73. *LAT*, 8, 13 February 1904.
74. *EPT*, 28 November 1905.
75. *EPT*, 17 February 1906.
76. *EPT*, 16 January 1909.
77. *EPT*, 21 January 1909; cited also in García, *Desert Immigrants*, 139.
78. Interview with Carmen Rangel Gonzáles, 3 May 1991, by Juliana Fahringer. In the possession of the author.
79. *EPT*, 15 February 1907.
80. *EPT*, 28 March 1905.
81. *EPT*, 23 February 1905; also cited in García, *Desert Immigrants*, 139.
82. F. Huerta, June 8, 1928, Chicago Notes, PTC.
83. Interview with Bert Corona, Los Angeles, California, November 10, 1989.
84. García, *Desert Immigrants*, 143–44.
85. "The Mexican 'Invaders' of El Paso," *The Survey*, (July 1916): 381.
86. *Socialist*, 15 March 1902.
87. Least Heat-Moon, *PrairyErth*, 235.
88. *Tetwiler's*, 13 June 1907.
89. I borrow this concept from Dirk Hoerder, "An Internationally Mobile Working Class and Its Press in North America: A Survey," in *The Immigrant Labor Press in North America, 1840s–1970s: An Annotated Bibliography*, ed. Dirk Hoerder and Christine Harzig (New York: Greenwood Press, 1987): 13.
90. This interpretation is consistent with the political generational approach argued by Mario T. García in *Mexican Americans: Leadership, Ideology, and Identity* (New Haven: Yale University Press, 1989): 1–22; see also Morton Keller, "Reflections on Politics and Generations in America," *Daedalus* 107 (Fall 1978): 123–35.

Notes to Chapter 6

1. The term "Mexican" is preferred here as opposed to the others for three reasons: the term Mexican historically encompasses

Indian and Hispano whereas the others do not; Euro-Americans lumped them all together as such in the evidence; and most importantly, Mexican immigrants or "migrants" (as is often argued) far outnumbered Indians and Hispanos on the railroad, especially in track work.

2. *Pacific Railway Commission*, 2589, cited in Tzu Kuei, "Chinese Workers," 1977, 117; see also Ronald T. Takaki, *Iron Cages: Race and Culture in Nineteenth Century America* (Seattle: University of Washington Press,1979) 171–93.
3. As discussed in chapter 2, only a few American Indians ever worked beyond one or two seasons on the tracks. Mexican immigrant labor ultimately replaced them as well.
4. For examples of Catholicism in New Mexico see Richard L. Nostrand, *The Hispano Homeland* (Norman: University of Oklahoma Press, 1992): 122–23; Howard Roberts Lamar, *The Far Southwest, 1846–1912: A Territorial History* (New Haven: Yale University Press, 1966): 200; for the Plains and Midwest see, *El Cosmopolita* (Kansas), 7 July 1917, 22 March 1919; *St. Paul Pioneer Press* (Minnesota), 19 December 1926; *Emporia Gazette* (Kansas), December 10, 1979. Francisco Arturo Rosales, "Mexicanos in Indiana Harbor During the 1920's: Prosperity and Depression," *Revista Chicano-Riquena* 4 (1976): 90–91; Michael Smith, *The Mexicans in Oklahoma* (Norman: University of Oklahoma Press, 1980): 58.
5. See Laird, "Argentine," 169; Harvey M. Chaldin, "Kinship Networks in the Migration Process" *International Migration Review* 7 (Summer 1973): 163–75; Eugene Litwack, "Geographic Mobility and Extended Family Cohesion," American Sociological Review (June 1960): 385–95; Ted J. Jitodai, "Migration and Kinship Contacts," *Pacific Sociological Review* 6 (Fall 1963): 49–55.
6. Charles Nordhoff, "The Human Side of Mexico," *Atlantic Monthly*, October 1919, 507.
7. *SFM*, November 1914, 27.
8. Licht, *Working for the Railroad*, 49.
9. "The Widowed Women of Santa Fe," in *Unequal Sisters*, ed. Ellen Carol Dubois and Vicki L. Ruiz (New York: Routledge, 1990): 38–39.
10. Deena González, 125–26 in manuscript. Book published as

Refusing the Favor: The Spanish-Mexican Women of Santa Fe, 1820-1880 (New York: Oxford University Press, 1999).

11. Ibid.
12. Ibid.
13. López interview.
14. It should be noted that the railroads employed white women as office workers early on, but they were typically paid less than their male counterparts. See Licht, *Working for the Railroad*, 214–15; Maurine Weiner Greenwald, "Women Workers and World War I: The American Railroad Industry, A Case Study," *Journal of Social History* 9 (Winter 1975): 154–77.
15. *Los Angeles Herald*, 1 June 1882.
16. Cuevas interview.
17. Ibid.
18. *SFM*, September 1910, 77.
19. Cuevas interview.
20. Santa Barbara *Morning Press*, 2 May 1900.
21. See chapter 4.
22. *SFM*, April 1911, 84–85.
23. Ibid.
24. Jonathan Hart, *Anarchism and the Mexican Working Class* (Austin: University of Texas Press, 1978). In some cases, mutualistas may have been modeled after Euro-American and African American mutual aid and secret societies.
25. Deutsch, *No Separate Refuge*, 26.
26. José Timoteo López, *La Historia de la Sociedad Proteccion Mutua de Trabajadores Unidos* (New York: Comet Press Books, 1958). According to Frederick C. Sanchez, there are no official documents for the nineteenth century. "S.P.M.D.T.U," 1.
27. *Witchcraft in the Southwest: Spanish and Indian Supernaturalism on the Rio Grande* (Lincoln: University of Nebraska Press, 1980): 94.
28. Elsie Clews Parsons, *Isleta, New Mexico* (Washington, D.C., 1932): 205.
29. Antonia Aparicio interview, 23 December 1990.
30. García, *Desert Immigrants*, 204; Arnoldo De Leon, *The Tejano Community, 1836–1900* (Albuquerque: University of New Mexico Press, 1982): 157–58; more recently Maria Herrera-Sobek has contributed to our understanding of the corrido and

traqueros; see her fine book, *Northward Bound: The Mexican Immigrant Experience in Ballad and Song* (Bloomington: University of Indiana Press, 1993): 34–63.

31. García, *Desert Immigrants*, 204.
32. Charles F. Lummis, *The Land of Poco Tiempo* (New York: Charles Scribners and Sons, 1925): 236–37.
33. Gamio, *Mexican Immigration*, 192–93.
34. Ibid., 84–86. Katy (K.T.) was the abbreviation for the Kansas and Texas Railroad.
35. Mexican Folkways, Corridos, Words and some Music, PTC.
36. Telephone interview with Alex Garcilazo, June 24, 1992.
37. Américo Paredes, *A Texas-Mexican Cancionero: Folksongs of the Lower Border* (Urbana: Illinois Press, 1976): 56–57.
38. García, *Desert Immigrants*, 204.
39. Gamio, *Mexican Immigration*, 96.
40. Some references to the "Mexican Problem" are Edwin F. Bamford, "The Mexican Casual Problem in the Southwest," *Journal of Applied Sociology* (July 1924): 363–71; Emory S. Bogardus, "The House-Court Problem," *American Journal of Sociology* (1917): 391–99; *Immigration and Race Attitudes* (Boston: D.C. Heath, 1928), passim.
41. See chapter 2 for discussion of the term "unskilled" as a category of analysis.
42. (Los Angeles: Wetzel Publishing Co., 1929): 31. Also cited in George J. Sánchez, "Becoming Mexican American: Ethnicity and Acculturation in Chicano Los Angeles, 1900–1943" (Ph.D. dissertation, Stanford University, 1989), 139.
43. George B. Hodgkin, "Making the Labor Camp Pay," *California Citrograph*, August 1921, 354; also cited in Gilbert González, "Labor and Community: The Camps of Mexican Citrus Pickers in Southern California," *Western Historical Quarterly* (August 1991): 289–312.
44. Evangeline Hymer, "A Study of the Social Attitudes of Adult Mexican Immigrants in Los Angeles and Vicinity" (M.A. thesis, University of Southern California, 1923), 51; also cited in George J. Sánchez, "'Go After the Women': Americanization and the Mexican Immigrant Woman, 1915–1929," in *Unequal Sisters*, ed. Ellen Carol Dubois and Vicki L. Ruiz (New York: Routledge, 1990): 260.
45. California, Commission of Immigration and Housing, *The*

Home Teacher, Immigrant Education Leaflet No. 5 (San Francisco: State Printing Office, 1916): 8. Also cited in Sánchez, "Becoming Mexican American," 138–39.

46. Bruce B. Bell, "Americanization in the Evening Schools," [Hammond] *Lake County Times*, 9 November 1926.
47. Ibid.; *El Cosmopolita*, 10 May 1919; also cited in Laird, "Argentine," 49.
48. Mary P. Ryan, *Cradle of the Middle Class: The Family in Oneida County, New York, 1790–1865* (Cambridge: Cambridge University Press, 1981), passim.
49. *Lake County Times*, 9 November 1926.
50. *SFM*, June 1919, 49.
51. Servando I. Esquivel, "The Immigrant from Mexico," *The Outlook*, 19 May 1920, 131.
52. Ibid.
53. *SFM*, February 1911, 88.
54. *Emporia Gazette*, 10 September 1979.
55. Chicago Notes, August 5, 1925, 73, PTC.
56. Hodgkin, 1921, 354.

notes to conclusion

1. Coatsworth, *Growth Against Development*, 7.
2. Parlee, "Railroad Unions," 445–46.
3. For example, see, *RFM*, March 1927, 94; May 1927, 208; October 1927, 429–31.
4. Barbara Driscoll, "The Railroad Bracero Program of World War II" (Ph.D. dissertation: University of Notre Dame, 1980).
5. Taylor, Arandas, 42.
6. Taylor, Arandas, 41; see also Bryan, "Mexican Immigrants," 728; Wollenberg, "Working on El Traque," 365.

BIBLIOGRAPHY

government publications

California, Commission of Immigration and Housing. *Report of an Experiment Made in Los Angeles in the Summer of 1917 for the Americanization of Foreign-Born Women.* Sacramento: State Printing Office, 1917.

Illinois, *Report of the Mayor's Commission on Unemployment.* Chicago: State Printing Office, March 1914.

U.S.Industrial Commission. *Reports of the Industrial Commission on Immigration,* 1901, rpt. Arno Press, 1970.

U.S.Congress. House of Representatives. *Annual Reports of the Department of the Interior for the Fiscal Year Ending June 30, 1898.* 55th Cong., 3rd Sess., Document no. 5. GPO. 1898.

U.S.Congress. *House of Representatives. Investigation of Labor Troubles in Missouri, Arkansas, Kansas, Texas, and Illinois.* 49th Cong., 2d. Sess., 3 March 1887.

U.S.Congress. *Senate. Immigration Commission, Immigrants in Industries Part 11: Slaughtering and Meat Packing. Part III: Slaughtering and Meat Packing in Kansas City, Kansas. 61st Cong. 2d. sess., 1910. Part 25: Japanese and Other Immigrant Races in the Pacific Coast and Rocky Mountain States. Part 1: Immigrant Laborers Employed by Steam Railway Companies in the Pacific Coast and Rocky Mountain States,* 61st Cong., 2d. sess., 1910.

U.S.Congress. Senate. Dillingham Commission. *Immigrants in Industries.* 61st Cong., 2nd sess., 1911, Doc. no. 633.

U.S.Department of Labor. *Mexican Labor in the United States*. By Victor S. Clark. Bulletin, No. 78. Washington, D.C.: GPO, 1908.

U.S. Department of Labor. *The Racial Problem Involved in Immigration from Latin America and the West Indies to the United States*. By Robert F. Foerster. Washington D.C.: GPO, 1925.

Manuscript Collections and Documents

Atchison, Topeka & Santa Fe Railway, Executive Files, Kansas State Historical Society, Topeka, Kansas.

Felipe Chavez Papers, State Records Center and Archives, Santa Fe, New Mexico.

José M. García, Typescript. Kansas State Historical Society. Topeka, Kansas.

Lee Haney. "Mexican Colony—Annual Fiesta," Typescript, U.S. Work Project Administration, Historical Records Survey.

Elfido López Sr. Collection, Colorado Historical Society. Denver, Colorado.

John M. Meade Manuscript. Kansas State Historical Society, Topeka, Kansas.

L. Bradford Prince, "White Caps" Investigation, 1892. State Records Center and Archives, Santa Fe, Mew Mexico.

Paul S. Taylor Collection, 74/187c, Bancroft Library, University of California, Berkeley.

U.S. Records of the National Mediation Board, Records of the Railway Labor Board. Case Files, 1920–26. Record Group 13, National Archives. Washington D.C.

U.S. Railroad Administration. Record Group 14, National Archives. Washington D.C.

U.S. Immigration and Naturalization Service. Record Group 85. National Archives. Washington, D.C.

U.S. Mediation and Conciliation Service. Case Files, Record Group 280. Washington National Records Center, Suitland, Md.

Oral History

Carmen Alvarado. July 28, 1980. Interviewed by Yolanda Coria. Lyon County.

Pascuál and Irma Aparicio. December 23, 1990. Interviewed by Jeffrey M. Garcilazo.

Bert Corona, November 10, 1989. Interviewed by Jeffrey M. Garcilazo.

José and Esperanza Cuevas. October 18, 1990. Interviewed by Jeffrey M. Garcilazo.

Lorenzo Delgado. July 23, 1980. Interviewed by Linda Hernandez. Lyon County.

José González and Lola Alcalá González. October 21, 1990. Interviewed by Jeffrey M. Garcilazo.

Carmen Rangel Gonzáles. May 3, 1991. Interviewed by Juliana Fahringer.

Pedro Ibarra. n.d., Laurie Bretz. KSHS.

Benjamin Lemos. May 19, 1991. Interviewed by Nancy Marshall.

Domingo and Antonia López. October 22, 1990. Interviewed by Jeffrey M. Garcilazo.

John López. August 1, 1980. Sandra Granado. Lyon County.

Michael López. August 1, 1980. Interviewed by Yolanda Coria. Lyon County.

John and Pascual Madrigal. April 29, 1980. Interviewed by Laurie Bretz. KSHS.

Armida Martínez. July 17, 1980. Interviewed by Sandra Granado.

Eugene Mendoza. July 3, 1980. Interviewed by Yolanda Coria. Lyon County.

James V. Mendoza. December 12, 1990. Interviewed by Jeffrey M. Garcilazo.

Adolph Oropeza. May 1980. Interviewed by Laurie Bretz. KSHS.

Quiroga, Melquiades. April 24, 1980. Interviewed by Laurie Bretz. KSHS.

Ramírez, Sehor y Sehora. July 28, 1980. Interviewed by Laurie Bretz. KSHS.

Simon Rodriguez. June 25, 1980. Interviewed by Sandra Granado. Lyon County.

Refugio Rochin Sr., April 9, 1992. Interviewed by Jeffrey M. Garcilazo.

Books

Acuña, Rodolfo. *Occupied America: The Chicano's Struggle Toward Liberation*. San Francisco: Canfield Press, 1972.

———. *Occupied America: A History of Chicanos.* 3rd. ed. New York: Harper and Row, 1988.

Adams, Ramon F. *The Language of the Railroader.* Norman: University of Oklahoma Press, 1977.

Allen, Ruth A. *The Great Southwestern Strike.* Austin: University of Texas Publication, 1942.

Allhands, James Lewellyn. *Gringo Builders.* Iowa City: Privately printed, 1931.

———. *Railroads to the Rio.* Salado, Texas: Anson Jones Press, 1960.

Almaguer, Tomás. *Racial Faultlines: The Historical Origins of White Supremacy in California.* Berkeley: University of California Press, 1994.

Armitage, Merle, and Edwin Corle, eds. *Operations Santa Fe: Atchison, Topeka, and Santa Fe Railway System.* New York: Duell, Sloan, and Pearce, 1948.

Athearn, Robert. *Rebel of the Rockies: The Denver and Rio Grande Western Railroad.* New Haven: Yale University Press, 1962.

Barrera, Mario. *Race and Class in the Southwest: A Theory of Racial Inequality.* Notre Dame: University of Notre Dame Press, 1979.

Bell, William Abraham. *New Tracks in North America.* London: Chapman and Hall, 1870.

Beveridge, W. H. *Unemployment: A Problem of Industry (1909 and 1930).* London: Longmans, Green and Co., 1931.

Bogardus, Emory S. *Immigration and Race Attitudes.* Boston: D.C. Heath and Company, 1928.

———. *The Mexican in the United States.* Los Angeles: University of Southern California Press, 1934.

Blaisdell, Lowell. *The Desert Revolution.* Madison: University of Wisconsin Press, 1962.

Botkin, B. A., and Alvin F. Harlow. *A Treasury of Railroad Folklore.* New York: Crown Publishers, 1953.

Bradley, Glenn Danford. *The Story of the Santa Fe.* Boston: Gorham Press, 1920.

Bradwin, Edmund W. *The Bunkhouse Man: A Study of Work and Pay in the Camps of Canada, 1903–1914.* New York: Columbia University Press, 1928.

Braverman, Harry. *Labor and Monopoly Capital: The Degradation of Work in the Twentieth Century.* New York: Monthly Review Press, 1974.

Brecher, Jeremy. *Strike!* Boston: Southend Press, 1987.

Brownlee, W. Elliot. *Dynamics of Ascent: A History of the American Economy*. New York: Alfred A. Knopf, 1974.

Bryant Jr., Keith L. *History of the Atchison, Topeka and Santa Fe Railway*. New York: Macmillan Publishing Co., 1974.

Camarillo, Albert. *Chicanos in a Changing Society: From Mexican Pueblos to American Barrios in Santa Barbara and Southern California, 1848–1930*. Cambridge: Harvard University Press, 1979.

———. *Chicanos in California: A History of Mexican Americans in California*. San Francisco: Boyd and Fraser, 1984.

Campa, Arthur L. *Hispanic Culture in the Southwest*. Norman: University of Oklahoma Press, 1993.

Chandler, Alfred D. *The Railroads: The Nation's First Big Business*. New York: Harcourt, Brace and World, Inc., 1965.

———. *The Visible Hand: The Managerial Revolution in American Business*. Cambridge: Harvard University Press, 1977.

Chavez, John R. *The Lost Land: The Chicano Image of the Southwest*. Albuquerque: University of New Mexico Press, 1984.

Chinn, Thomas W. *A History of the Chinese in California*. San Francisco: Chinese Historical Society of America, 1969.

Chiu, Ping. *Chinese Labor in California, 1850–1880*. Madison: University of Wisconsin Press, 1963.

Coatsworth, John. *Growth Against Development: The Economic Impact of Railroads in Porfirian Mexico*. De Kalb: Northern Illinois University Press, 1981.

Cockroft, James D. *Intellectual Precursors of the Mexican Revolution, 1900–1913*. Austin: University of Texas Press, 1971.

Coolidge, Mary Roberts. *Chinese Immigration*. New York: H. Holt & Company, 1909.

Commons, John R., et al., *History of Labor in the United States*. 4 volumes. New York: Macmillan Company, 1921–35.

Conkling, Roscoe and William D. Shipman. *The Central Pacific Railroad Company in Equitable Account with the United States, Growing out of Subsidy Bonds in Aid of Construction*. New York: Henry Bessey, 1887.

Corwin, Arthur F. *Immigrants—and Immigrants: Perspectives on Mexican Labor Migration to the United States*. Westport: Greenwood Press, 1978.

Crump, Spencer. *Ride the Big Red Cars: How the Trolleys Helped Build Southern California.* 3d. ed. Costa Mesa: Trans-Anglo Books, 1970.

Cumberland, Charles C. *Mexican Revolution: Genesis Under Madero.* Austin: University of Texas Press, 1952.

De Leon, Arnoldo. *They Called Them Greasers: Anglo Attitudes Toward Mexicans in Texas, 1821–1900.* Austin: University of Texas Press, 1983.

———. *The Tejano Community, 1836–1900.* Albuquerque: University of New Mexico Press, 1982.

———. *Mexican Americans in Texas: A Brief History.* Arlington Heights, Illinois: Harlan Davidson, 1993.

Deutsch, Sarah. *No Separate Refuge: Culture, Class, and Gender on an Anglo-Hispanic Frontier in the American Southwest, 1880–1940.* New York: Oxford University Press, 1987.

Deverell, William F. *Railroad Crossing: Californians and the Railroad, 1850–1910.* Berkeley: University of California Press, 1994.

Duke, Donald, and Stan Kistler. *Santa Fe...Steel Rails Through California.* San Marino, California: Pacific Railroad Publications, 1964.

Dubofsky, Melvyn. *We Shall Be All: A History of the IWW.* New York: Quadrangle, 1969.

Ducker, James H. *Men of the Steel Rails: Workers on the Atchison, Topeka and Santa Fe Railroad, 1869–1900.* Lincoln: University of Nebraska Press, 1983.

Dunbar-Ortiz, Roxanne. *Roots of Resistance: Land Tenure in New Mexico, 1680–1980.* Los Angeles: Chicano Studies Research Center Publications, 1980.

Ellis, Pearl Idelia. *Americanization Through Homemaking.* Los Angeles: Wetzel Publishing Co., 1929.

Fink, Leon. *Workingmen's Democracy.* Urbana: University of Illinois Press, 1983.

Fishlow, Albert. *American Railroads and the Transformation of the Ante-Bellum Economy.* Cambridge: Harvard University Press, 1965.

Foster, William Z. *Pages from a Worker's Life.* New York: International Publishers, [1939] 1978.

Galarza, Ernesto. *Merchants of Labor: The Mexican Bracero Story.* Charlotte: McNally and Loftin, 1964.

Gamio, Manuel. *Mexican Immigration to the United States: A Study of Human Migration and Adjustment.* 1930; rpt. New York: Dover, 1971.

———. *The Life Story of the Mexican Immigrant: Autobiographical Documents.* 1931; rpt. New York: Dover Publications, 1971.

García, Mario T. *Desert Immigrants: The Mexicans of El Paso, 1880–1920.* New Haven: Yale University Press, 1981.

———. *Mexican Americans: Leadership, Ideology, and Identity, 1930–1960.* New Haven: Yale University Press, 1989.

Garlock, Jonathan. *Guide to the Local Assemblies of the Knights of Labor.* Westport: Greenwood Press, 1982.

Glaab, Charles N. *Kansas City and the Railroads: Community Policy in the Growth of a Regional Metropolis.* Madison: State Historical Society of Wisconsin, 1962.

Gómez-Quiñones, Juan. *Development of the Mexican Working Class North of the Rio Bravo: Work and Culture Among Laborers and Artisans, 1600–1900.* Los Angeles: Chicano Studies Research Center Publications, 1982.

———. *Sembradores.* Los Angeles: Chicano Studies Center Publications, 1977

Gonzalez, Gilbert G. *Labor and Community: Mexican Citrus Worker Villages in a Southern California County, 1900–1950.* Urbana: University of Illinois Press, 1994.

Graham, Richard, ed., *The Idea of Race in Latin America, 1870–1940.* Austin: University of Texas Press, 1990.

Grant, Blanche C. *When Old Trails Were New: The Story of Taos.* Glorieta, NM: Rio Grande Press, 1983.

Grant, H. Roger, ed. *Brownie the Boomer: The Life of Charles P. Brown, an American Railroader.* DeKalb: University of Northern Illinois Press, 1991.

Griswold del Castillo, Richard. *The Los Angeles Barrio, 1850–1890: A Social History.* Berkeley: University of California Press, 1979.

———. *The Treaty of Guadalupe Hidalgo: A Legacy of Conflict.* Norman: Oklahoma University Press, 1990.

Greenwald, Maurine Weiner. *Women, War and Work: The Impact of World War I on Women Workers in the United States.* Westport: Greenwood Press, 1980.

Guerin-Gonzales, Camille. *Mexican Workers and American Dreams: Immigration, Repatriation, and California Farm Labor, 1900–1939.* New Brunswick: Rutgers University Press, 1994.

Gurley, F. G. *New Mexico and the Santa Fe Railway*. New York: The Newcomen Society in North America, 1950.

Gutiérrez, Ramón A. *"When Jesús Came the Corn Mothers Went Away": Marriage, Sexuality, and Power in New Mexico, 1500–1846*. Stanford: Stanford University Press, 1991.

Gutman, Herbert G. *Work, Culture, and Society in Industrializing America: Essays in American Working Class and Social History*. New York: Vintage Books, 1977.

———. *Power and Culture: Essays on the American Working Class*. New York: The New Press, 1987.

Hall, Thomas D. *Social Change in the Southwest, 1350–1880*. Lawrence: University of Kansas Press, 1989.

Harper, Allan, Andrew Cordova, and Kalervo Oberg. *Man and Resources in the Middle Rio Grande Valley*. Albuquerque: University of New Mexico Press, 1943.

Hart, John M. *Anarchism and the Mexican Working Class, 1860–1931*. Austin: University of Texas Press, 1978.

———. *Revolutionary Mexico: The Coming Process of the Mexican Revolution*. Berkeley: University of California Press, 1989.

Herrera-Sobek, Maria. *Northward Bound: The Mexican Immigrant Experience in Ballad and Song*. Bloomington: Indiana University Press, 1993.

Hertel, D. W. *History of the Brotherhood of Maintenance of Way Employees: Its Birth and Growth, 1887–1955*. Washington D.C.: Ransdell Inc., 1955.

Hoerder, Dirk. Ed. *"Struggle the Hard Battle": Essays on Working-Class Immigrants*. DeKalb: Northern Illinois University Press. 1986.

Hufford, Charles H. *The Social and Economic Effects of Mexican Migration into Texas*. San Francisco: R&E Associates [1925] 1971.

Kern, Robert, ed. *Labor in New Mexico: Unions, Strikes, and Social History Since 1881*. Albuquerque: University of New Mexico Press, 1983.

Korman, Gerd. *Industrialization, Immigrants, and Americanizers: The View from Milwaukee, 1866–1921*. Madison: State Historical Society of Wisconsin, 1967.

Lamar, Howard Roberts. *The Far Southwest, 1846–1912: A Territorial History*. New Haven: Yale University Press, 1966.

Lamphere, Louise, Patricia Zavella, and Felipe Gonzales with Peter Evans. *Sunbelt Working Mothers: Reconciling Family and Factory*. Ithaca: Cornell University Press, 1993.

Least Heat-Moon, William. *PrairyErth*. Boston: Houghton Mifflin Company, 1991.

Lescohier, Don Divance. *The Labor Market*. New York: Macmillan, 1919.

Licht, Walter. *Working for the Railroad: The Organization of Work in the Nineteenth Century*. Princeton, New Jersey: Princeton University Press, 1983.

Limerick, Patricia Nelson. *The Legacy of Conquest: The Unbroken Past of the American West*. New York: W.W. Norton and Company, 1987.

Lummis, Charles F. *The Land of Poco Tiempo*. New York: Charles Scribner's Sons, 1925.

MacLachlan, Colin M. *Anarchism and the Mexican Revolution: The Political Trials of Ricardo Flores Magón in the United States*. Berkeley: University of California Press, 1991.

Marks, Carole. *Farewell—We're Good and Gone: The Great Black Migration*. Bloomington: Indiana University Press, 1989.

Marshall, James. *Santa Fe: The Railroad that Built an Empire*. New York: Random House, 1945.

Martínez, Oscar J. *Border Boom Town: Ciudad Juárez since 1848*. Austin: University of Texas Press, 1978.

Mayer, Vicente V. Jr. ed. *Utah: A Hispanic History*. Salt Lake City: University of Utah Press, 1975.

McCombs, Vernon Monroe. *From Over the Border: A Study of the Mexicans in the United States*. New York: Council of Women for Home Missions and Missionary Education Movement, 1925.

McLean, Robert N. *The Northern Mexican*. New York: Home Missions Council, 1930, Reprinted R&E Associates [1971].

McWilliams, Carey. *North From Mexico: The Spanish-Speaking People of the United States*. New York: Greenwood Press, 1968.

Meier, Matt S. and Feliciano Rivera. *The Chicanos: A History of Mexican Americans*. New York: Hill and Wang, 1972.

Meinig, D. W. *Southwest: Three Peoples in Geographical Change 1600–1970*. New York: Oxford University Press, 1971.

Mexicans in California. *Report of Governor C. C. Young's Mexican Fact-Finding Committee*. 1930; rpt. San Francisco: R&E Associates, 1970.

Montejano, David. *Anglos and Mexicans in the Making of Texas, 1836–1986*. Austin: University of Texas Press, 1988.

Montgomery, David. *Fall of the House of Labor: The Workplace, the State, and American Labor Activism, 1865–1925*. Cambridge: Cambridge University Press, 1987.

Nadeau, Remi. *City Makers: The Story of Southern California's First Boom, 1868–76*. Los Angeles: Trans-Anglo Books, 1965.

Nash, Gerald D. *The American West in the Twentieth Century: A Short History of an Urban Oasis*. Englewood Cliffs, New Jersey: Prentice Hall, 1973.

Nee, Victor G. and Bret de Barry Nee. *Longtime Californ': A Documentary Study of an American Chinatown*. Stanford: Stanford University Press, 1986.

Nostrand, Richard L. *The Hispano Homeland*. Norman: University of Oklahoma Press, 1992.

Paredes, Américo. *A Texas-Mexican Cancionero: Folksongs of the Lower Border*. Urbana: University of Illinois Press, 1976.

Parsons, Elsie C. *Isleta, New Mexico*. Washington D.C.: Bureau of Ethnology, 1932.

Peña, Manuel. *The Texas-Mexican Conjunto: History of a Working-Class Music*. Austin: University of Texas, 1985.

Perry, Louis B. and Richard S. Perry. *A History of the Los Angeles Labor Movement, 1911–1941*. Berkeley: University of California Press, 1963.

Pletcher, David M. *Rails, Mines, and Progress: Seven American Promoters in Mexico, 1867–1911*. Ithaca: Cornell University Press, 1958.

Portes, Alejandro, Manuel Castells and Lauren A. Benton. *The Informal Economy: Studies in Advanced and Less Developed Countries*. Baltimore: John Hopkins University Press, 1989.

Quastler, I. E. *Railroads of Lawrence, Kansas: 1854–1900*. Lawrence: Coronado Press, 1979.

Quinn, Anthony. *The Original Sin—A Self-Portrait*. Boston: Little Brown and Company, 1972.

Quirk, Robert E. *An Affair of Honor: Woodrow Wilson and the Occupation of Veracruz*. New York: W.W. Norton & Co., 1962.

Raat, W. Dirk. *Revoltosos*. College Station: Texas A&M University Press, 1981.

Rachleff, Peter. *Black Labor in Richmond, 1865–1890*. Urbana: University of Illinois Press, 1989.

Reisler, Mark. *By the Sweat of their Brow: Mexican Immigrant Labor in the United States, 1900–1940*. Westport, Connecticut: Greenwood Press, 1976.

Rickard, Thomas A. *A History of American Mining*.New York: Johnson Reprints, 1932.

Richardson, Reed. *The Locomotive Engineer, 1863–1963: A Century of Railway Labor Relations and Work Rules*. Ann Arbor: University of Michigan Press, 1963.

Riegel, Robert Edgar. *The Story of the Western Railroads*. New York: The Macmillan Company, 1926.

Risher, Howard W. *The Negro in the Railroad Industry*. Philadelphia: University of Pennsylvania Press, 1971.

Rodea, Marcelo N. *Historia del Movimiento Obrero Ferrocarrilero en Mexico (1890–1943)*. Mexico: n.p., 1944.

Romo, Ricardo. *East Los Angeles: History of a Barrio*. Austin: University of Texas Press, 1983.

Rosenbaum, Robert J. *Mexicano Resistance in the Southwest: "The Sacred Right of Self-Preservation"*. Austin: University of Texas Press, 1981.

Ruíz, Ramón E. *The Great Rebellion: Mexico, 1905–1924*. New York: W. W. Norton, 1980.

Ruíz, Vicki L. *Cannery Women Cannery Lives: Mexican Women, Unionization, and the California Food Processing Industry, 1930–1950*. Albuquerque: University of New Mexico Press, 1987.

Rubel, Arthur J. *Across the Tracks: Mexican-Americans in a Texas City*. Austin: University of Texas Press, 1966.

Ryan, Mary P. *Cradle of the Middle Class: The Family in Oneida County, New York, 1790–1865*. Cambridge: Cambridge University Press, 1981.

Sabin, Edwin L. *Building the Pacific Railway*. Philadelphia: J. B. Lippincott Co., 1919.

Salas, Elizabeth. *Soldaderas in the Mexican Military: Myth and History*. Austin: University of Texas Press, 1990.

Saldívar, Ramón. *Chicano Narrative: The Dialectics of Difference*. Madison: University of Wisconsin Press, 1990.

Sánchez, George J. *Becoming Mexican American: Ethnicity, Culture and Identity in Chicano Los Angeles, 1900–1945*. New York: Oxford University Press, 1993.

Saxton, Alexander. *The Indispensable Enemy: Labor and the Anti-Chinese Movement in California*. Berkeley: University of California Press, 1971.

Schlebecker, John T. *Cattle Raising on the Plains 1900–1961*. Lincoln: University of Nebraska Press, 1963.

Seidman, Joel. *The Brotherhood of Railroad Trainmen: The Internal Political Life of a National Union*. New York: John Wiley and Sons Inc., 1962.

Sheridan, Thomas E. *Los Tucsonenses: The Mexican Community in Tucson, 1854–1941*. Tucson: University of Arizona Press, 1992.

Simmons, Marc. *Witchcraft in the Southwest: Spanish and Indian Supernaturalism on the Rio Grande*. Lincoln: University of Nebraska Press, 1980.

Smith, Michael M. *The Mexicans in Oklahoma*. Norman: University of Oklahoma Press, 1980.

Stimson, Grace Heilman. *Rise of the Labor Movement in Los Angeles*. Berkeley: University of California Press, 1955.

Stover, John F. *American Railroads*. Chicago: University of Chicago Press, 1961.

Stromquist, Shelton. *A Generation of Boomers: The Pattern of Railroad Labor Conflict in Nineteenth-Century America*. Urbana: University of Illinois Press, 1987.

Sunseri, Alvin R. *Seeds of Discord: New Mexico in the Aftermath of the American Conquest, 1846–1861*. Chicago: Nelson-Hall, 1979.

Swadesh, Frances Leon. *Los Primeros Pobladores: Hispanic Americans of the Ute Frontier*. Notre Dame: University of Notre Dame Press, 1974.

Takaki, Ronald T. *Iron Cages: Race and Culture in Nineteenth-Century America*. Seattle: University of Washington Press, 1979.

Taylor, Fredrick Winslow. *Shop Management*. New York: Harper and Brothers, 1903.

———. *The Principles of Scientific Management*. New York: Harper & Row, [1911] 1967.

Taylor, Paul S. *Mexican Labor in the United States*. 7 volumes. Berkeley: University of California Press, 1929–1934.

———. *A Spanish-Mexican Peasant Community: Arandas in Jalisco, Mexico*. Berkeley: University of California Press, 1933.

Thompson, E. P. *The Making of the English Working Class*. New York: Vintage Books, 1966.

Tratman, E. E. Russell. *Railway Track and Track Work*. New York: McGraw-Hill, 1909.

Turner, John Kenneth. *Barbarous Mexico*. Austin: University of Texas Press, [1910] 1969.

Valdés, Dennis Nodin. *Al Norte: Agricultural Workers in the Great Lakes Region, 1917–1970*. Austin: University of Texas Press, 1991.

Vargas, Zaragosa. *Proletarians of the North: A History of Mexican Industrial Workers in Detroit and the Midwest, 1917–1933*. Berkeley: University of California Press, 1993.

Venegas, Daniel. *Las Aventuras de Don Chipote: O Cuando Los Pericos Mamen*. Mexico: Secretaría de Educación Pública, [1928] 1984.

Waters, L. L. *Steel Rails to Santa Fe*. Lawrence: University of Kansas Press, 1950.

Way, Peter. *Common Labour: Workers and the Digging of the North American Canals, 1780–1860*. Cambridge: Cambridge University Press, 1993.

White, Owen. *Out of the Desert: The Historical Romance of El Paso*. El Paso: McMath Co., 1923.

White, Richard. *It's Your Misfortune and None of My Own: A New History of the American West*. Norman: University of Oklahoma Press, 1991.

Wilentz, Sean. *Chants Democratic: New York City and the Rise of the American Working Class, 1788–1850*. New York: Oxford University Press, 1950.

Womack, John Jr. *Zapata and the Mexican Revolution*. New York: Vintage, 1969.

Zamora, Emilio. *The World of the Mexican Worker in Texas*. College Station: Texas A&M University Press, 1993.

Zavella, Patricia. *Women's Work and Chicano Families: Cannery Workers of the Santa Clara Valley*. Ithaca: Cornell University Press, 1987.

Articles

Almaguer, Tomás. "Chicano Men: A Cartography of Homosexual Identity and Behavior." *Differences: A Journal of Feminist Cultural Studies* 3 (1991): 75–100.

———. "Ideological Distortions in Recent Chicano Historiography: the Internal Model and Chicano Historical Interpretation." *Aztlan* (1989): 7–28.

———."Class, Race, and Chicano Oppression." *Socialist Revolution* (July–September 1975): 71–99.

———. "Toward the Study of Chicano Colonialism." *Aztlan* (Spring 1971): 7–21.

Alvarez, Rodolfo. "The Psycho-historical and Socioeconomic Development of the Chicano Community in the United States." *Social Science Quarterly* 53 (1973): 920–42.

Armitage, Sue. Theresa Banfield, and Sarah Jacobus. "Black Women and their Communities in Colorado." *Frontiers* 2 (Summer 1977): 36-40.

Arroyo, Luis Leobardo. "The State of Chicano Labor History 1970–1980." In *Chicanos and the Social Sciences: A Decade of Research and Development (1970–1980)*. Isidro D. Ortiz, ed. (Santa Barbara: Center for Chicano Studies, 1983): 1–8.

Bamford, Edwin F. "Industrialization and the Mexican Casual." *Proceedings of the Southwestern Political and Social Science Association* (Austin: March 1924): 251–64.

Barth, Gunther. "Chinese Sojourners in the West: The Coming." *Southern California Quarterly* 46 (March 1964): 55–67.

Brody, David. "The Old History and the New: In Search of an American Working Class." *Labor History* 20 (1979): 111–26.

Bryan, Samuel. "Mexican Immigrants in the United States." *The Survey* 28 (September 1912): 726–30.

Carroll, Bernice A. "On Mary Beard's Woman as Force in History." In Bernice A. Carroll, ed., *Liberating Women's History: Theoretical and Critical Essays* (Urbana: University of Illinois Press, 1976): 26–41.

Cassity, Michael J. "Modernization and Social Crisis: The Knights of Labor and a Midwest Community, 1885–1886." *Journal of American History* 66 (June 1979): 41-61.

Castillo, Pedro. "The Making of the Mexican Working Class in the United States: Los Angeles, California: 1880–1920." In *El Trabajo Y Los Trabajadores En La Historia De Mexico*. Elsa Cecilia Frost, et al., eds. El Colegio de Mexico y University of Arizona Press, 1982, 506–17.

Choldin, Harvey M. "Kinship Networks in the Migration Process." *International Migration Review* 7 (Summer 1973): 163–75.

Clark, Victor S. "Mexican Labor in the United States." *Bulletin of the Department of Labor* 78 (September 1908): 469–74.

Coatsworth, John Henry. "Railroads and the Concentration of Landownership in the Early Porfiriato." *Hispanic American Historical Review* 54 (February 1974): 48–71.

Corwin, Arthur F. "Causes of Mexican Emigration to the United States: A Summary View." *Perspectives in American History* 7 (1973): 557–635.

Culhane, John. "Hero Street USA." *Reader's Digest* (May 1985): 81–86.

Davis, Colin J. "Bitter Conflict: the 1922 Railroad Shopmen's Strike." *Labor History* 33 (Fall 1992): 433–55.

Eggert, Gerald G. "A Missed Alternative: Federal Court Arbitration of Railway Disputes, 1877–1895." *Labor History* 7 (Fall 1966): 287–306.

Estrada, Leobardo F., F. Chris García, Reynaldo Flores Macias and Lionel Maldonado, "Chicanos in the United States: A History of Exploitation and Resistance." *Daedalus* (Spring 1981): 103–31.

Fuentes, Carlos. "Hispanic USA: The Mirror of the Other." *The Nation* (March 30, 1992): 409.

Fuller, Roden. "Occupations of the Mexican-Born Population of Texas, New Mexico and Arizona, 1900–1920." *Journal of the American Statistical Association* (March 1928): 64–67.

Galarza, Ernesto. "Mexicans in the Southwest: A Culture in Process." In *Plural Society in the Southwest.* Edward Spicer and Raymond H. Thompson, eds. (Albuquerque: University of New Mexico Press, 1975): 261–97.

García, Juan R. "Midwest Mexicanos in the 1920s: Issues, Questions, and Directions." *Social Science Journal* 19 (April 1982): 89–99.

García, Mario T. "Americanization and the Mexican Immigrant, 1880-1939." *The Journal of Ethnic Studies* 6 (1978): 19-34.

———. "The Chicana in American History: The Mexican Women of El Paso, 1880–1920—A Case Study." *Pacific Historical Review* 49 (1980): 315–38.

———. "La Familia: The Mexican Immigrant Family, 1900–1930." In *Work, Family, Sex Roles, Language: The National Association for Chicano Studies, Selected Papers 1979*. Mario Barrera, Alberto Camarillo, and Francisco Hernandez. eds. (Berkeley: Tonatiuh-Quinto Sol International, 1980): 117–39.

Gómez-Quiñones, Juan. "On Culture." *Revista Chicano-Riqueña* 5 (1977): 29–47.

———. "Mexican Immigration to the United States and the Internationalization of Labor, 1848-1980: An overview."In *Mexican Immigrant Workers in the U.S.* Antonio Rios-Bustamante, ed. (Los Angeles: Chicano Studies Research Center Publications, 1981):13–34.

Graziosi, Andrea. "Common Laborers, Unskilled Workers: 1880–1915." *Labor History* 22 (Fall 1981): 512–44.

Greenwald, Maurine Weiner. "Women Workers and World War I: The American Railroad Industry, A Case Study." *Journal of Social History* 9 (Winter 1975): 154–77.

Greever, William S. "Railway Development in the Southwest." *New Mexico Historical Review* 32 (1957): 151–203.

Griego, Andrew. "Rebuilding the California Southern Railroad: The Personal Account of a Chinese Labor Contractor, 1884." *Journal of San Diego History* 25 (1979): 324–37.

Hobsbawm, Eric J. "Afterword: Working Classes and Nations." In *Labor Migration in Atlantic Economies: The European and North American Working-Classes During the Period of Industrialization.* Dirk Hoerder. ed. (Westport: Greenwood Press, 1985): 435–47.

Hoerder, Dirk. "An Internationally Mobile Working Class and Its Press in North America: A Survey." In *The Immigrant Labor Press in North America, 1840s–1970s: An Annotated Bibliography.* Dirk Hoerder and Christine Harzig. eds. (New York: Greenwood Press, 1987).

Jameson, Elizabeth. "Women as Workers, Women as Civilizers: True Womanhood in the American West." In *The Women's West.* Susan Armitage and Elizabeth Jameson, eds. (Norman: University of Oklahoma Press, 1987): 145–64.

Jitodai, Ted T. "Migration and Kinship Contacts." *Pacific Sociological Review* 6 (Fall 1963): 49–55.

Knight, Alan. "Racism, Revolution, and Indigenismo: Mexico, 1910–1940." In *The Idea of Race in Latin America, 1870–1940.* Richard Graham, ed. (Austin: University of Texas Press, 1990): 71–113.

Knowlton, Clark S. "Land-Grant Problems among the State's Spanish-Americans." *New Mexico Business* (June 1967): 1–13.

Krueger, Thomas A. "American Labor Historiography, Old and New: A Review Essay." *Journal of Social History* 4 (Spring, 1971): 277–285.

Lamar, Howard. "From Bondage to Contract: Ethnic Labor in the American West, 1600–1890." In *The Countryside in the Age of Capitalist Transformation: Essays in the Social History of Rural America*. Steven Hahn and Jonathan Prude, eds. (Chapel Hill: University of North Carolina Press, 1985): 293–324.

Lesley, Lewis B. "The Entrance of the Santa Fé into California." *Pacific Historical Review* 8 (1939): 89–96.

Marks, Carole. "Lines of Communication, Recruitment Mechanisms, and the Great Migration of 1916–1918." *Social Problems* 31 (October 1983): 73–83.

Marston, Helen D. "Mexican Traits." *The Survey* (August 1920): 562–64.

Martinez, Oscar J. "On the Size of the Chicano Population: New Estimates, 1850–1900." *Aztlán* 6 (Spring 1975): 43–67.

McLean, Robert N. "Mexican Workers in the United States." *Proceedings of the National Conference on Social Work*, San Francisco (1929): 531–38.

Meléndez, A. Gabriel. "Camellando en los Estamos Sumidos." *Bilingual Review* 13 (September–December 1986): 87–93.

"The Mexican 'Invaders' of El Paso." *The Survey* (July 1916): 380–82.

Miller, Richard Ulric. "American Railroad Unions and the National Railways of Mexico: An Exercise in Nineteenth-Century Proletarian Manifest Destiny." *Labor History* 15 (1974): 239–60.

Montgomery, David. "Thinking About American Workers in the 1920s." *International Labor and Working-Class History* 32 (Fall, 1987): 4–22.

———. "To Study the People: The American Working Class." *Labor History* 21 (Fall 1980): 485–512.

———. "Nationalism, American Patriotism, and Class Consciousness among Immigrant Workers in the United States in the Epoch of World War I." In *"Struggle the Hard Battle": Essays on Working-Class Immigrants*. Dirk Hoerder, ed. (DeKalb: Northern Illinois University Press. 1986): 327–51.

Oppenheimer, Robert. "Acculturation or Assimilation: Mexican Immigrants in Kansas, 1900 to World War II." *Western Historical Quarterly* 16 (October 1985): 429–48.

Pannell, W. W. "Mexican Workers in the Southwest." *International Socialist Review* 16 (September 1916): 168–69.

Parlee, Lorena. "The Impact of United States Railroad Unions on Organized Labor and Government Policy in Mexico (1880–1911)." *Hispanic American Historical Review* 64 (August 1984): 443–75.

Pascoe, Peggy. "Race, Gender, and Intercultural Relations: The Case of Interracial Marriage." *Frontiers* 12 (1991–1992): 1–4.

Peters, Kurt M. "Watering the Flower: The Laguna Pueblo and the Santa Fe Railroad, 1880–1943." In *American Indian Wage Labor*. Martha C. Knack and Alice Littlefield, eds. (Norman: University of Oklahoma, 1994).

Ray, William W. "Crusade or Civil War?: The Pullman Strike in California." *California History* 58 (Spring 1979): 20–37.

Romo, Ricardo. "The Urbanization of Southwestern Chicanos in the Early 20th Century." In *New Directions in Chicano Scholarship*. Ricardo Romo and Raymund Paredes, eds. (Santa Barbara: Center for Chicano Studies, 1984): 183–207.

Rosales, Francisco Arturo. "Mexicanos in Indiana Harbor During the 1920's: Prosperity and Depression." *Revista Chicano-Riqueña* 4 (1976): 88–98.

———. "Mexican Immigrant Experience in the Urban Midwest: East Chicago, Indiana, 1919–1945." *Indiana Magazine of History* (December 1981): 333–57.

Sanchez, Frederick C. "A History of the S.P.M.D.T.U" *San Luis Valley Historian* 3 (Winter 1971): 1–16.

Sanchez, George F. "'Go After the Women': Americanization and the Mexican Immigrant Woman, 1915–1929." In *Unequal Sisters*. Ellen Carol Dubois and Vicki Ruiz, eds. (New York: Routledge, 1990): 250-63.

Saldivar, José David. "Américo Paredes and Decolonization." In *Cultures of Imperialism*. Amy Kaplan and Donald E. Pease, eds. (Durham: Duke University Press, 1993): 292–311.

Saxton, Alexander. "The Army of Canton in the High Sierra." *Pacific Historical Review* 35 (1966): 141–52.

Schlesinger, Andrew Bancroft. "Las Gorras Blancas, 1889–1891." *Journal of Mexican-American History* (Spring 1971): 87–143.

Smith, Michael. "The Mexican Immigrant Press Beyond the Borderlands." *Great Plains Quarterly* (Spring 1990): 71–85.

———. "Mexicans in Kansas City: The First Generation, 1900–1920." *Perspectives on Mexican American Studies* (1989): 29–57.

———. "Beyond the Borderlands: Mexican Labor in the Central Plains, 1900–1930." *Great Plains Quarterly* (Fall 1981): 239–51.

Stromquist, Shelton. "Enginemen and Shopmen, Technological Change and the Organization of Labor in an Era of Railroad Expansion." *Labor History* 24 (Fall 1983): 48–99.

Taylor, Paul S. "Some Aspects of Mexican Immigration." *Journal of Political Economy* 38 (October 1930): 609–15.

———. "Note on Streams of Mexican Migration." *American Journal of Sociology* 36 (1930–31): 287–88.

Tugwell, Rexford G., "The Casual of the Woods." *The Survey* (July 1920): 472–73.

Van Arsdale, Jonathan. "Railroads in New Mexico." *Research* 2 (December 1939): 3–16.

Vargas, Zaragosa. "Life and Community in the 'Wonderful City of the Magic Motor': Mexican Immigrants in 1920s Detroit." *Michigan Historical Review* 15 (Spring 1989): 45–68.

Wollenberg, Charles. "Working on El Traque: The Pacific Electric Strike of 1903." *California Historical Quarterly* 42 (August 1973): 358–69.

Zamora, Emilio. "Chicano Socialist Labor Activity in Texas 1900–1920." *Aztlán* (1976): 221–36.

Zavella, Patricia. "Feminist Insider Dilemmas: Constructing Ethnic Identity with 'Chicanal Informants." *Frontiers* 13 (1993): 53–76.

Zieger, Robert. "From Hostility to Moderation: Railroad Labor Policy in the 1920s." *Labor History* 9 (Winter 1968): 23–38.

Dissertations, M.A. Theses, and Unpublished Papers

Berger, Sister Wilhelmina. "The Beginnings of Catholicity in Kansas City, Kansas." (M.A. thesis, Creighton University 1934).

Castillo, Pedro G. "Making of a Mexican Barrio: Los Angeles, 1890–1920." (Ph.D. dissertation. University of California, Santa Barbara 1979).

Braun, Frances. "A Survey of the American-Mexicans in Topeka, Kansas." (M.A. thesis. Kansas State Teachers College 1970).

Case, Theresa Ann. "Class as Culture: Gender and Race in the 1886 Southwest Strike." Paper presented at the Southwest Labor Studies Conference. Santa Cruz, California. April 29, 1994.

Coatsworth, John Henry. "The Impact of the Railroads on the Economic Development of Mexico, 1877–1910." (Ph.D. dissertation. University of Wisconsin 1973).

Davis, Colin J. "The Contours of Race and Ethnicity: The 1922 National Railroad Strike." Paper presented at the Southwest Labor Studies Conference. Santa Cruz, California. April 29, 1994.

Escobar, Edward J. "Chicano Protest and the Law: Law Enforcement Responses to Chicano Activism in Los Angeles, 1850–1936." (Ph.D. dissertation. University of California, Riverside 1983).

Franco, Hector. "The Mexican People in the State of Kansas." (Ph.D. dissertation. University of Wichita 1950).

Graves, Carl Russell. "Scientific Management and the Santa Fe Railway Shopmen of Topeka, Kansas, 1900–1925." (Ph.D. dissertation. Harvard University 1980).

Haas, Mary Lisbeth. "The Barrios of Santa Ana: Community, Class, and Urbanization, 1850–1947." (Ph.D. dissertation. University of California, Irvine 1985).

Laird, Judith Fincher. "Argentine Kansas: The Evolution of a Mexican-American Community, 1905–1940." (Ph.D. dissertation. University of Kansas 1975).

Lightner, David Lee. "Labor on the Illinois Central Railroad, 1852–1900." (Ph.D. dissertation. Cornell University 1975).

Lin, Paul King-Chang. "Voluntary Kinship and Voluntary Association in a Mexican-American Community." (M.A. thesis, University of Kansas, 1963).

Mines, Cynthia. "Riding the Rails to Kansas: The Mexican Immigrants." (M.A. thesis, McPherson College, 1980).

Park, Joseph F. "The History of Mexican Labor in Arizona During the Territorial Period." (M.A. thesis, University of Arizona, 1961).

Parlee, Lorena M. "Porfirio Diaz, Railroads, and Development in Northern Mexico: A Study of Government Policy Toward the Central and Nacional Railroads, 1876–1910." (Ph.D. dissertation. University of California, San Diego, 1981).

Peters, Kurt M. "Watering the Flower" The Laguna Pueblo and the Santa Fe Railroad, 1880–1943." Paper presented at the Southwest Labor Studies Conference. Santa Cruz, California. April 29, 1994.

Ramírez, Socorro M. "A Survey of Mexicans in Emporia, Kansas." (M.A. thesis, Emporia, Kansas State Teachers College 1942).

Ricart, Domingo. "Just Across the Tracks: Report on a Survey of Five Communities in the State of Kansas, Emporia, Florence, Newton, Wichita, Hutchinson." (M.A. thesis. Lawrence, University of Kansas 1950).

Rutter, Larry G. "Mexican Americans in Kansas: A Survey and Social Mobility Study, 1900–1977." (M.A. thesis. Kansas State University Manhattan 1972).

Sánchez, George J. "Becoming Mexican American: Ethnicity and Acculturation in Chicano Los Angeles, 1900–1943." (Ph.D. dissertation. University of California, Los Angeles 1989).
Seligmann Jr., Gustav Leonard. "The El Paso and Northeastern Railroad System and Its Economic Influence in New Mexico." (M.A. thesis. New Mexico State College 1958).
Swan, Robert A. Jr. "The Ethnic Heritage of Topeka: Immigrant Beginnings." (Topeka: Institute of Comparative Ethnic Studies, 1974).
Vargas, Zaragosa. "Mexican Auto Workers at Ford Motor Company, 1918–1933." (Ph.D. dissertation. University of Michigan 1984).
White, W. Thomas. "A History of Railroad Workers in the Pacific Northwest, 1883–1934." (Ph.D. dissertation. University of Washington 1981)
Yen, Tzu-Kuei. "Chinese Workers and the First Transcontinental Railroad of the United States." (Ph.D. dissertation. St. Johns University 1977).

Newspapers, Periodicals and Trade Magazines

Albuquerque Evening Citizen.
Albuquerque Morning Journal.
American Labor Union Journal.
Arizona Daily Star.
ArizonaWeekly Star.
Arizona Sentinel (Arizona City).
Arizona Silver Belt (Globe).
Colton Chronicle.
Colton Courier.
El Cosmopolita (Kansas City, Missouri).
El Grito.
El Paso Herald.
El Paso Times.
Fort Sumner Review (New Mexico).
Fronteriza.
Industrial Worker.
International Socialist Review.
La Prensa (San Antonio, Texas).
Labor Press.

La Voz del Pueblo (Las Vegas, New Mexico).
Las Vegas Democrat.
Las Vegas Daily Optic.
Los Angeles Daily Herald.
Los Angeles Record.
Los Angeles Socialist.
Los Angeles Times.
Nation.
Outlook.
Pacific Electric Magazine.
Presscott Weekly Courier (Arizona).
Railway Age Gazette.
Railroad Association Magazine.
Railway Employees Journal.
Railway Engineering & Maintenance of Way.
Railway Maintenance Engineer.
Rebelde (Los Angeles).
Regeneración (Los Angeles).
Rock Island Employees Magazine.
San Bernardino Index.
San Bernardino Sun.
San Diego Union.
Santa Barbara Morning Press.
Santa Fe Employee's Magazine.
Santa Fe New Mexican.
Solidarity.
Survey.
Tetwiler's.
Tombstone Epitaph.
Tucson Citizen.
Topeka Daily Capital.
Voice of Labor.
World's Work.

Miscellaneous

The Iron Road, 60 min., distributed by PBS video, Alexandria, Virginia, 1990.

index

www.ingramcontent.com/pod-product-compliance
Lightning Source LLC
Jackson TN
JSHW071230010626
103187JS00009B/95
9781574416275